Network Flow Models
and Applications

The Author

Er. Shruti Singh did M.Tech (Computer Science and Engineering) from Shri Ramswaroop Memorial University (SRMU) Lucknow and B.Tech (Computer Science and Engineering) from Gautam Buddh Technical University (GBTU), Lucknow. She worked as an Assistant Professor at Himalayan Institute of Technology and Management, Lucknow. She has published number of research papers in national and international journals of repute and also presented papers in several conferences. She has received various honours and awards in the field of computer science and engineering.

Network Flow Models and Applications

Author
Shruti Singh

Kruger Brentt
Publishers

2024

Kruger Brentt Publishers UK. LTD.
Company Number 9728962

Regd. Office: 68 St Margarets Road, Edgware, Middlesex HA8 9UU

Library of Congress Cataloging-in-Publication Data

ISBN 978-1-78715-014-0 (Hardbound)

For information on all our publications visit our website at http://krugerbrentt.com/

Preface

Network flow problems are central problems in operations research, computer science, and engineering and they arise in many real world applications. Everywhere we look in our daily lives, networks are apparent. Electrical and power networks bring lighting and entertainment into our homes. Telephone networks permit us to communicate with each other almost effortlessly within our local communities and across regional and international borders. National highway systems, rail networks, and airline service networks provide us with the means to cross great geographical distances to accomplish our work, to see our loved ones, and to visit new places and enjoy new experiences. Manufacturing and distribution networks give us access to life's essential food stock and to consumer products. And computer networks, such as airline reservation systems, have changed the way we share information and conduct our business and personal lives. In all of these problem domains, and in many more, we wish to move some entity (electricity, a consumer product, a person or a vehicle, a message) from one point to another in an underlying network, and to do so as efficiently as possible, both to provide good service to the users of the network and to use the underlying (and typically expensive) transmission facilities effectively.

Network flows is a problem domain that lies at the cusp between several fields of inquiry, including applied mathematics, computer science, engineering, management, and operations research. The field has a rich and long tradition, tracing its roots back to the work of Gustav Kirchhof and other early pioneers of electrical engineering and mechanics who first systematically analyzed electrical circuits. This early work set the foundations of many of the key ideas of network flow theory and established networks (graphs) as useful mathematical objects for representing many physical systems.

The present book discusses various fundamental concepts in the subject that are updated within the newest development in the field. This book covers network and models, multi-terminal multipath flows, synthesis of networks, network simplex method, applications of network flow, problem solving with network flow and project scheduling with resource constraints. This is an excellent book for network flow courses of the graduate and under graduate levels, professionals working with network flow, optimization and network programming. In spite of the best efforts, it is possible that some errors may have occurred into the compilation and editing of the book. Further queries, constructive suggestions and criticisms for improvement of the book are always welcome and shall be thankfully acknowledged. Last but not the least, it is a pleasure for us to extend our sincere thanks to Kruger Brentt Publisher, London (UK) and his team members for his keen interest shown in preparation and publishing of this book so efficiently and promptly.

Er. Shruti Singh

Contents

1 | Network and Models

1.1 Introduction

A network is a set of objects called nodes or vertices that are connected together. The connections between the nodes are called edges or links. In mathematics, networks are often referred to as graphs which must be distinguished from an alternative use of the graph to mean a graph of a function.

If the edges in a network are directed, *i.e.*, pointing in only one direction, the network is called a directed network or digraph. When drawing a directed network, the edges are typically drawn as arrows indicating the direction, as illustrated in the Figure 1.1.

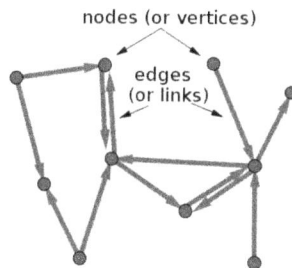

Figure 1.1: A Directed Network with 10 Nodes (or vertices) and 13 Edges (or links).

If all edges are bidirectional, or undirected, the network is an undirected network or graph as shown in Figure 1.2.

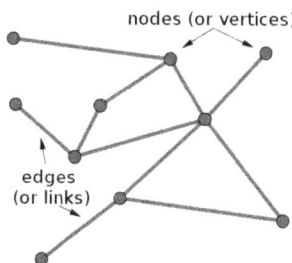

Figure 1.2: An Undirected Network with 10 Nodes (or vertices) and 11 Edges (or links).

According to Euler, the four land areas be named as A, B, C, and D as shown in Figure 1.3 which serve as four points with seven bridges as seven lines joining these points.

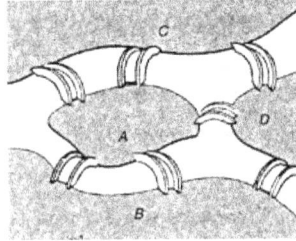

Figure 1.3: Network Graph.

According to him, the area (A) can be reached by five bridges as shown in figure which shows that there are five lines meeting at point A. The three lines from point D represent three bridges, *etc.* Such type of diagram is called as network graph, or simply a *network*. We see that Euler was concerned not with the size and shape of the bridges and land regions but rather with how the bridges were connected.

A network is a collection of points known as vertices, and a collection of lines is known as arcs that connects these points. A network is traversable, which shows that on tracing each arc exactly once by beginning at some point and not lifting the pencil from the paper. The problem of crossing each bridge exactly once reduces to one of traversing the network representing these bridges (Figure 1.4).

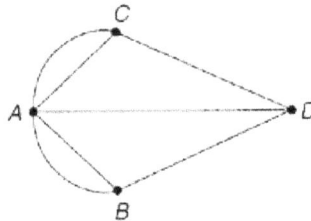

Figure 1.4: Network.

As per Euler, a network is traversable depending upon the number of odd vertices. In a network, if there are odd number of arcs at point A, then point A is called as odd vertex. If the number of arcs meeting at a point is *even*, then the point is known as even vertex. Euler found that the only traversable networks are those that have either no odd vertices or exactly have two odd vertices. As per Konigsberg network with four odd vertices, it is not traversable. So, it is impossible to take a walk over the bridges and can cross each bridge only once.

In a network graph, one can formally define an undirected graph as G=(N,E), consisting of the set N of nodes and the set E of edges, which are unordered pairs of elements of N. The formal definition of a directed graph is similar, but the only difference is that the set E contains ordered pairs of elements of N.

Networks can represent all sorts of systems in the real world. For example, Internet is also a network where nodes are computers or other devices and the edges are physical connections between the devices. World Wide Web is a huge network where the pages are nodes and links are the edges (Figure 1.5).

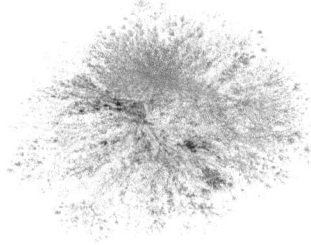

Figure 1.5: Network of Connections between Devices within the Internet.

Examples of a network include:

> ➢ social networks of acquaintances
> ➢ networks of publications linked by citations
> ➢ transportation networks
> ➢ metabolic networks
> ➢ communication networks

Example 1.1

Which of the following networks are traversable?

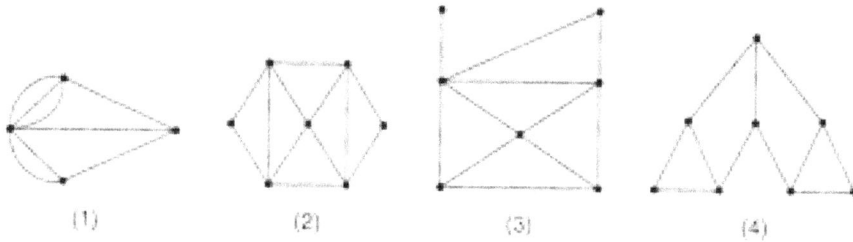

(1)   (2)   (3)   (4)

In the above figures:
Network 1 has two odd vertices, so it is traversable
Network 2 has no odd vertices, so it is traversable
Networks 3 and 4 have four and six odd vertices, respectively, so they are not traversable

Example 1.2

In the figures, each of these networks has two odd vertices and is traversable. Show a beginning point and an ending point for each network. Form a conjecture about the beginning and ending points of networks with exactly two odd vertices.

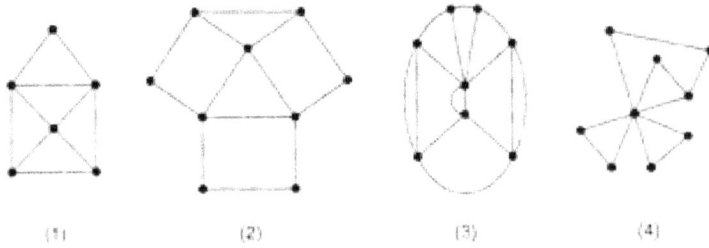

In case of conjecture, there will be one odd vertex which will be at the starting point and the other odd vertex will be the ending point. Consider the conjecture as shown in above example. Since the network is traversed, only two arcs are used each time when we pass through a vertex point. One arc will pass on arriving at the point and one at leaving the point. The can be only one way when there will odd vertex in a traversable network, when the vertex is at the beginning point or at the ending point.

Example 1.3

In the figures shown, the following networks have no odd vertices as all the vertices are even. Find at least two different beginning points for traversing each network. Form a conjecture about traversing networks with no odd vertices.

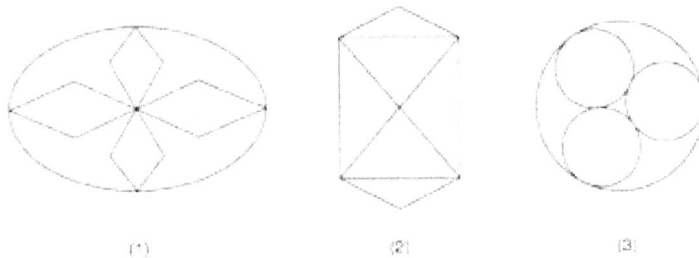

In case of conjecture in a traversing a network, all even vertices are present at the beginning point that may be any vertex, and at the ending point with same vertex.

The conjecture shown in example 3 seems reasonable. Since the arcs occur in pairs at each vertex, beginning at a vertex will always require returning to that vertex. The facts as highlighted in examples 1 to 3 are summarized as:

1.1.1 Traversable Networks

1. A network with exactly two odd vertices is traversable. Either odd vertex may be the beginning point, and the other odd vertex is the ending point.

2. A network with no odd vertices is traversable. Any vertex may be the beginning point, and the same vertex will also be the ending point.

3. A network with more than two odd vertices is not traversable.

Example 1.4

Each of the networks shown has two odd vertices and is traversable. Show a beginning point and an ending point for each. What is always true about the beginning and ending points of networks with exactly two odd vertices?

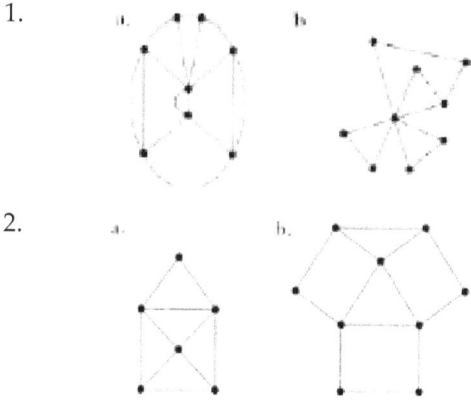

1.

2.

Which of the networks are traversable? Trace each traversable network on a separate piece of paper. Mark a path with arrows, and indicate the beginning and ending points.

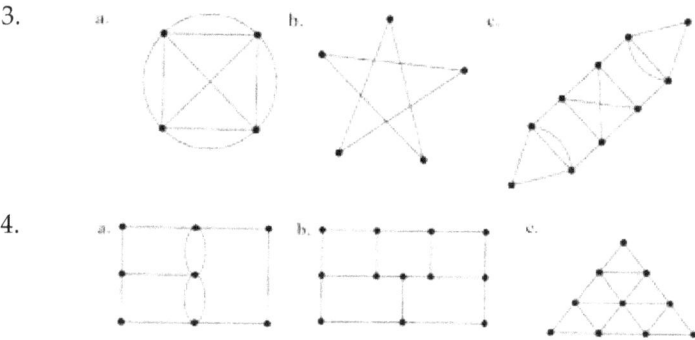

3.

4.

The vertices and edges of polyhedra are three-dimensional networks.

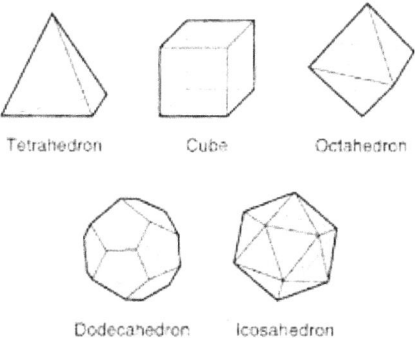

Tetrahedron   Cube   Octahedron

Dodecahedron   Icosahedron

1.1.2 Matrices Associated with Network Graphs

It is seen that a graph is formed by vertices and edges connecting the vertices.

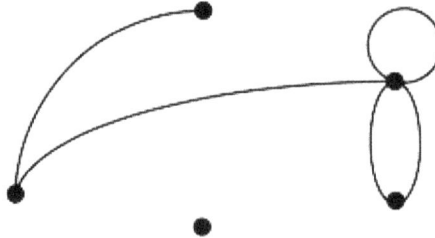

It is seen that a graph is a pair of sets (V, E), where V is the set of vertices and E is the set of edges, formed by pairs of vertices. E is a multiset, in other words, its elements can occur more than once so that every element has a multiplicity. If we label the vertices with letters a, b, c, . . . or v_1, v_2, . . . or numbers by 1, 2, . . ., then the elements will be labeled as V. If we label the vertices as:

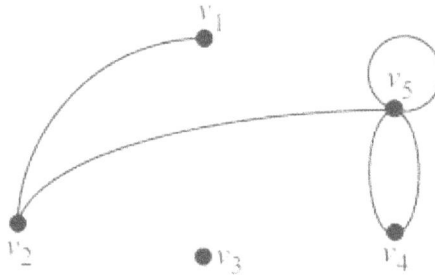

Figure 1.6: Labeling of Vertices.

So, V = {v1, . . ., v5} for the vertices and E = {(v1, v2),(v2, v5),(v5, v5),(v5, v4),(v5, v4)} for the edges, so the two edges (u, v) and (v, u) are the same. In other words, the pair is not ordered. If we label the edges as:

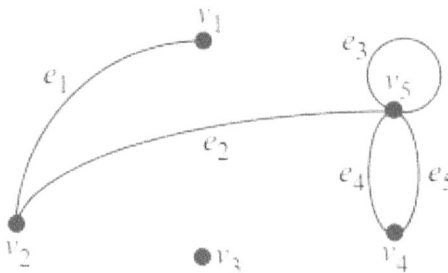

Figure 1.7: Labeling of Vertices.

So E = {e1, . . ., e5}.

We can say that:

1. The two vertices u and v are end vertices of the edge (u, v).
2. Edges that have the same end vertices are parallel.
3. An edge of the form (v, v) is a loop.
4. A graph is simple if it has no parallel edges or loops.
5. A graph with no edges (*i.e.* E is empty) is empty.
6. A graph with no vertices (*i.e.* V and E are empty) is a null graph.
7. A graph with only one vertex is trivial.
8. Edges are adjacent if they share a common end vertex.
9. Two vertices u and v are adjacent if they are connected by an edge, in other words, (u, v) is an edge.
10. The degree of the vertex v, written as d(v), is the number of edges with v as an end vertex. By convention, we count a loop twice and parallel edges contribute separately.
11. A pendant vertex is a vertex whose degree is 1.
12. An edge that has a pendant vertex as an end vertex is a pendant edge.
13. An isolated vertex is a vertex whose degree is 0.

In the Figure 1.7:

➢ v4 and v5 are end vertices of e5.
➢ e4 and e5 are parallel.
➢ e3 is a loop.
➢ The graph is not simple.
➢ e1 and e2 are adjacent
➢ v1 and v2 are adjacent.
➢ The degree of v1 is 1 so it is a pendant vertex.
➢ e1 is a pendant edge.
➢ The degree of v5 is 5.
➢ The degree of v4 is 2.
➢ The degree of v3 is 0 so it is an isolated vertex.

In the future, we will label graphs with letters, for example:

G = (V, E).

The minimum degree of the vertices in a graph G is denoted $\delta(G)$ (= 0 if there is an isolated vertex in G). Similarly, we write $\Delta(G)$ as the maximum degree of vertices in G.

Remember

> ➢ Define graph $G = (V, E)$ by explaining a pair of sets:
>> 1. V = set of **vertices**
>> 2. E = set of **edges**

Edges:

> ➢ Every edge is defined by pair of vertices
>
> ➢ Edge connects the vertices which define it
>
> ➢ In certain cases, vertices will be same

Vertices:

> ➢ Vertices are **nodes**
>
> ➢ Shows vertices with labels

Representation:

> ➢ Show vertices with circles or containing a label
>
> ➢ Show edges with lines among circles

Example 1.5

> ➢ V = {A,B,C,D}
>
> ➢ E = {(A,B),(A,C),(A,D),(B,D),(C,D)}

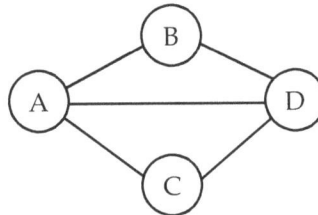

1.2 Types of Network Graphs

When one tries to model systems such as those mentioned above, one quickly realizes that the simple network model with identical nodes and edges cannot describe important features of real networks. One problem is the edges in this simplest network model are undirected. However, in the World Wide Web, for example, the links between pages are directed.

There are many common kinds of graphs

> ➢ Weighted or unweighted
>
> ➢ Directed or undirected
>
> ➢ Cyclic or acyclic

Graphs can be classified by weight of the edges:

> ➢ Weighted graph: Edges with weight that shows cost of traversing like distances between cities

> ➢ Unweighted graph: Edges without weight that show connections like course prereqs

Graphs can be classified based on edges direction:

> ➢ Undirected Graphs: Every edge can be traversed in either direction

> ➢ Directed Graphs: Every edge can be traversed in particular direction

1.2.1 Undirected Network Graph

In undirected graph, each edge is a two-element subset of V. A simple undirected graph contains no duplicate edges and no loops. A graph with more than one edge between the same two vertices is called multigraph. Many times, when we say graph, it means a simple undirected graph (Figure 1.8).

Figure 1.8: Undirected Graph.

Simple undirected graphs also correspond to relations, with the restriction that the relation must be irreflexive (no loops) and symmetric or undirected edges. This also gives a representation of undirected graphs as directed graphs, where the edges of the directed graph always appear in pairs going in opposite directions.

Further edges of a graph are assumed to be unordered pairs of vertices. Sometimes we say undirected graph to emphasize this point. In an undirected graph, we write edges using curly braces to denote unordered pairs. For example, an undirected edge {2,3} from vertex 2 to vertex 3 is the same thing as an undirected edge {3,2} from vertex 3 to vertex 2 (Figure 1.9).

This type of edge is found in protein-protein interaction networks (PPINs). The relationship between the nodes is a simple connection, without a given 'flow' implied, since the evidence behind the relationship only tells us that A binds B (Figure 1.10).

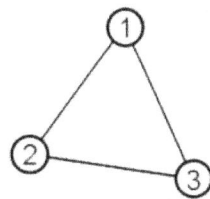

Undirected graph (V_1, E_1)

$V_1 = \{1,2,3\}$

$E_1 = \{\{1,2\},\{2,3\},\{3,1\}\}$

Figure 1.10: Undirected Edge.

Figure 1.10: Protein-protein Interaction Network.

Example

In a undirected graph G: V = {w,x,.y,z} and E = {(w,x), (x, w), (x, y), (y, x), (w, z), (z, w), (z, x), (x, z), (z, y), (y, z)}, so on verifying relation E as symmetric, the graph is shown.

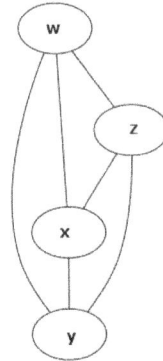

Example

If V = {Program, end, computer, math} and E = {(Program, end), (end, Program), (Program, computer), (computer, Program), (Program, math), (math, Program), (math, end), (end, math) are called self-loops.

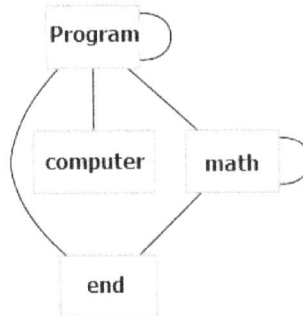

Adjacency and Degree

There are two ways of representing a graph inside a computer: adjacency list or a adjacency matrix. Two vertices u and v in an undirected graph are called adjacent (or neighbors) in G if e = {u,v} is an edge of G. The edge e is said to connect u and v. The vertices u and v are called the endpoints of e.

The degree of a vertex in an undirected graph is the number of edges that are incident with (or connect) it, except that a loop at a vertex contributes twice to the degree. The degree of the vertex v is denoted by deg(v).

In/Out Degree

Let G be a directed graph with vertex set V and edge set E.

The set of incoming edges of a vertex v are all those edges whose arrows point into v:

Incoming (v) = {(u,v) | (u,v) ε E}

In-Degree

In-Degree For any vertex v ε V, the in-degree of v is the number of incoming edges into v.

Similarly, we can define outgoing edges for a given vertex v:

Outgoing (x) = {(x,y) | (x,y) ε E}

Out-Degree

For any vertex v ε V, the out-degree of v is the number of outgoing edges out of v.

Note For a undirected graph, the set of incoming edges is the same as the set of out-going edges for any vertex

Degree

The degree of a vertex v in an undirected graph is the number of (we can say either incoming or outgoing) edges that are incident on v.

Note that the concepts of in-degree and out-degree coincide with that of degree for an undirected graph.

Example

How many edges are there in a graph with ten vertices each of degree 6?

Solution

Since the sum of the degrees of the vertices is 6 x 10 = 60

This shows that 2e = 60

Hence, e = 30

Example

In a undirected graph G, if V = {p,q,r,s} and E = {(p,q), (q,p), (q,r), (r,q), (p,r), (rr,p), (p,s), (s,p), (q,s), (s,q), (s,r), (r,s)}, then what will be the sequence of degree?

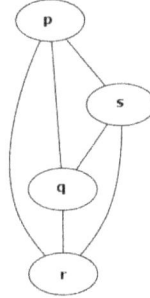

Now,

Node	Degree
p	3
q	3
r	3
s	3

The degree sequence is 3,3,3,3.

1.2.2 Directed Network Graph

In a directed graph, the two directions are counted as being distinct directed edges. In an directed graph, we write edges using parentheses to denote ordered pairs. For example, edge (2,3) is directed from 2 to 3, which is different than the directed edge (3,2) from 3 to 2. Directed graphs are drawn with arrowheads on the links, as shown in Figure 1.11.

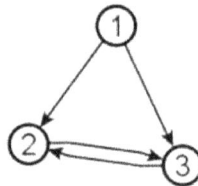

Directed graph (V.,E.)

$V_2 = \{1,2,3\}$

$E_2 = \{(1.2),(2.3),(3.2),(1.3)\}$

Figure 1.11: Directed Edge.

Directed edge is a kind of connection found in metabolic or gene regulation networks. There is a clear flow of signal implied and the network can be organised hierarchically (Figure 1.12).

Figure 1.12: Gene Regulation Network.

In a directed graph or digraph, each element of E is an ordered pair, and we think of edges as arrows from a source, head, or initial vertex to a sink, tail, or terminal vertex; each of these two vertices is called an endpoint of the edge. A directed graph is simple if there is at most one edge from one vertex to another. A directed graph that has multiple edges from some vertex u to some other vertex v is called a directed multigraph (Figure 1.13).

Figure 1.13: Directed Multigraph.

Example

If V = {a,b,c,d} and E = {(a,b), (b,c), (a,c), (d,b), (d,a), (d,c)}, then the graph of this is shown as:

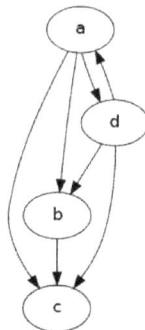

Example

If V = {Program, end, computer, math} and E = {(Program, Program), (Program, end), (Program, computer), (Program, math), (math, math), (math, Program)}, then edge (Program, Program) and the edge (math, math) are called self-loops, since they point from a vertex to itself.

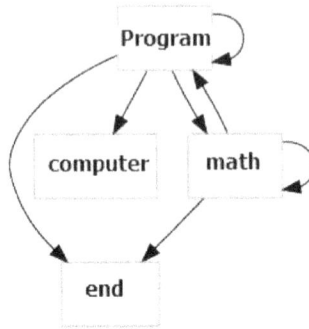

1.2.3 Cyclic Graphs

A cyclic graph is a directed graph with at least one cycle. A cycle is a path along the directed edges from a vertex to itself. Cyclic graphs contain cycles where it is possible to find a path from starting vertex traversing a set of unique edges and ending up back at starting vertex (Figure 1.14).

Cyclic graph can be a directed cycle graph where all the edges are oriented in similar direction. In this graph, set of edges that has at least one edge or arc from every directed cycle is known as feedback arc set. Similarly, a set of vertices containing at least one vertex from each directed cycle is called a feedback vertex set. A directed cycle graph has uniform in-degree 1 and uniform out-degree 1.

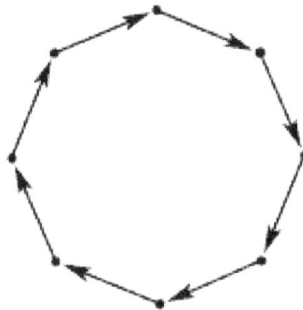

Figure 1.14: Directed Cycle Graph.

1.2.4 Directed Acyclic Graphs

A cyclic graph is a graph without cycles. In this, while following the graph from node to node, same node cannot be visited twice (Figure 1.15).

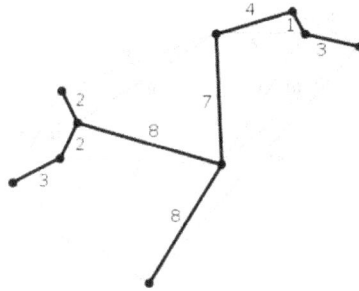

Figure 1.15: A Cyclic Graph.

Further it is observed that a connected acyclic graph is known as a tree where one or more of tree branches gets disconnected which results in forest as shown in Figure 1.16.

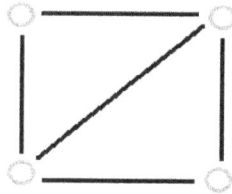

Figure 1.16: Forest Graph.

To support ability so as to push and pull change sets among many instances of similar repository, a special designed structure for showing various versions of things known as Directed Acyclic Graph or DAG is designed that is more expressive as compared to purely linear model (Figure 1.17).

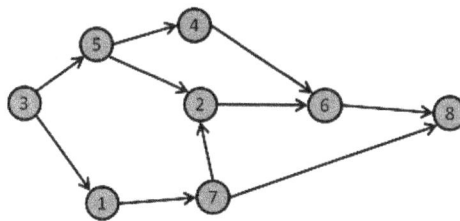

Figure 1.17: Directed Acyclic Graph.

The parts of above graph are:

- ➢ Underlying set for the Vertices set is Integer.
- ➢ Vertices set = {1,2,3,4,5,6,7,8}
- ➢ Edge set = {(1,7),(2,6),(3,1),(3,5),(4,6),(5,4),(5,2),(6,8),(7,2),(7,8)}

It is observed that in a directed acyclic graph, there results a topological ordering where the nodes are ordered so that, starting node has low value as compared to ending node. In DAG, there appears a unique topological ordering when it contains a directed path of the nodes.

Also, history of everything in repository is modeled as DAG where a second generation tools modeled history of repository as a line which involves sequence of versions, one after the other as shown in Figure 1.18.

Figure 1.18: Repository History as a Line.

Linear history for its simplicity is liked by the people as it gives an unambiguous answer to the question about its latest version. There exists a problem with the linear model as new version can only be committed when it was based on latest version which happens commonly.

DAG being a data structure from computer science that modeled many problems, so it has following elements:

> Nodes: Every node shows certain object or piece of data, so in DVCS, a node shows one revision of entire repository tree.

> Directed edges: A directed edge from one node to other shows certain relationship among two nodes. Here arrows in DAG will not form a cycle.

> Root node: One of nodes will have no parents which serves as root of DAG.

> Leaf nodes: One or more of nodes will have no children, so these are leaves or leaf nodes.

1.2.5 Weighted Network Graph

Directed or undirected edges can also have weight or a quantitative value associated with them. This is used to depict concepts such a reliability of an interaction, the quantitative expression change that a gene induces over another or even how closely related two genes are in terms of sequence similarity. Edges can also be weighted by their centrality values or several other topological parameters (Figure 1.19).

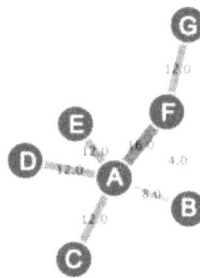

Figure 1.19: Genes Relationship Network.

Consider a graph of 4 nodes as shown in Figure 1.20.

Figure 1.20: Weighted Graph.

As you can see each edge has a weight/cost assigned to it. Suppose we need to go from vertex 1 to vertex 3. There are 3 paths.

1 -> 2 -> 3
1 -> 3
1 -> 4 -> 3
Total cost of 1 -> 2 -> 3 will be (1 + 2) = 3 units.
Total cost of 1 -> 3 will be 1 units.
Total cost of 1 -> 4 -> 3 will be (3 + 2) = 5 units.

1.2.6 Hypergraphs

In a hypergraph, the edges are called hyperedges that results as arbitrary nonempty sets of vertices. A k-hypergraph is one in which all such hyperedges connected exactly k vertices, hence an ordinary graph is thus a 2-hypergraph. Hypergraphs are usually drawn by representing each hyperedge as a closed curve containing its members as (Figure 1.21):

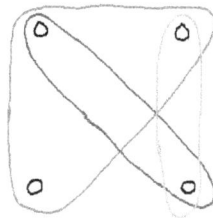

Figure 1.21: Hypergraphs.

Hypergraphs are not used commonly as it is always possible to show a hypergraph by a bipartite graph.

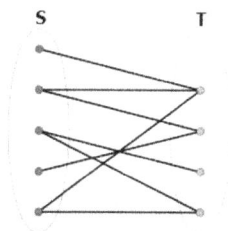

In a bipartite graph, the vertex set can be partitioned into two subsets S and T, such that every edge connects a vertex in S with a vertex in T. To represent a hypergraph H as a bipartite graph, we simply represent the vertices of H as vertices in S and the hyperedges of H as vertices in T, and put in an edge (s,t) whenever s is a member of the hyperedge t in H.

Example:

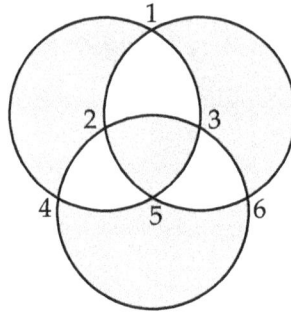

V(M) = {1, 2, 3, 4, 5, 6}

E(M) = {12, 24, 14, 13, 36, 16, 23, 25, 35, 45, 56, 46}

F(M) = {124, 136, 235, 456, 123, 245, 356, 146}

V(H) = {1, 2, 3, 4, 5, 6}

E(H) = {124, 136, 235, 456}

F(H) = {123, 245, 356, 146}

1.2.7 Vertex Labeled Graphs

In a labeled graph, each vertex is labeled with some data in addition to the data that identifies the vertex. Only the indentifying data is present in the pair in the Edge set. This is similar to the key, satellite data distinction for sorting (Figure 1.22).

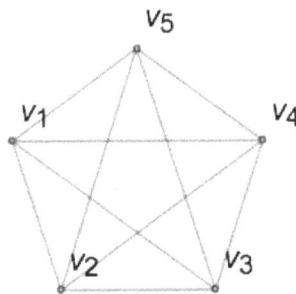

Figure 1.22: Vertex Label Graph.

Consider following parts:

> ➢ Underlying set for the keys of the Vertices set is the integers.

> ➢ Underlying set for the satellite data is Color.

> ➢ Vertices set = {(2,Blue),(4,Blue),(5,Red),(7,Green),(6,Red),(3,Yellow)}

> ➢ Edge set = {(2,4),(4,5),(5,7),(7,6),(6,2),(4,3),(3,7)}

1.2.8 Edge Labeled Graphs

Edge labeled graph is a graph where the edges are associated with labels. One can indicate this be making the Edge set be a set of triples. Thus if (u,v,X) is in the edge set, then there is an edge from u to v with label X. Edge labeled graphs are usually drawn with the labels drawn adjacent to the arcs specifying the edges (Figure 1.23).

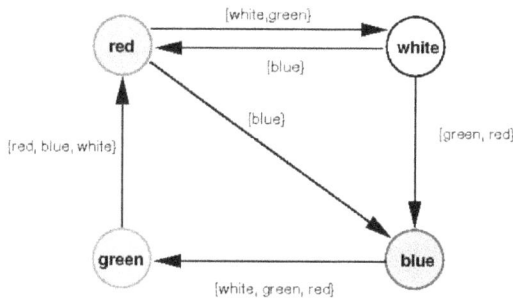

Figure 1.23: Edge Labeled Graph.

It has following parts:

> ➢ Underlying set for the the Vertices set is Color.

> ➢ Underlying set for the edge labels is sets of Color.

> ➢ Vertices set = {Red,Green,Blue,White}

> ➢ Edge set = {(red, white, {white, green}), (white, red, {blue}), (white, blue, {green, red}), (red, blue, {blue}), (green, red, {red, blue, white}), (blue, green, {white,green,red})}

1.2.9 Connected and Disconnected Graphs

Connected Graph

A graph is connected when there is a path between every pair of vertices. In this type of graph, there are no unreachable vertices. Here, every vertex to other vertex, there result certain path which can be traversed. It is known as connectivity of a graph. A graph with multiple disconnected vertices and edges is said to be disconnected.

A connected graph 'G' may have at most (n–2) cut vertices. If G is a connected graph and a vertex V ∈ G is a cut vertex of G, the G-V results in disconnected graph. Removing a cut vertex from a graph breaks it in to two or more graphs (Figure 1.24).

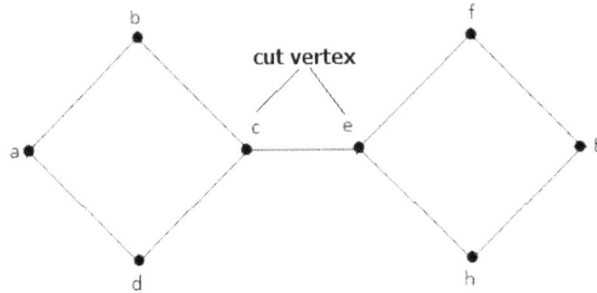

Figure 1.24: Cut Vertices.

Disconnected Graph

A graph is said to be a disconnected graph, if there exist two nodes in G such that no path in G has those nodes as endpoints.

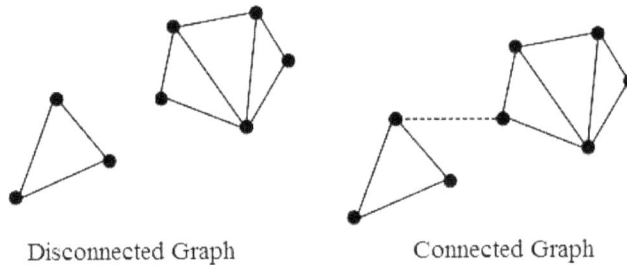

Disconnected Graph   Connected Graph

1.2.10 Planar and Non-Planar Graphs

When a connected graph can be drawn without any edges crossing, it is called a planar. Graph A is planar since no link is overlapping with another. Graph B is non-planar since many links are overlapping. It is observed that the links of graph B cannot be reconfigured in a manner that would make it planar (Figure 1.25).

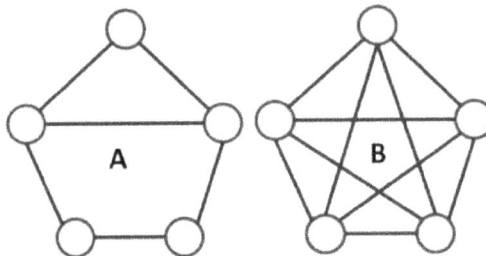

Figure 1.25: Planar and Non Planar Graph.

A graph is planar if it can be drawn in a plane in such a way that no two edges meet each other except at a vertex to which they are incident. Graphs below are planar, since they can be drawn in the plane without edges crossing.

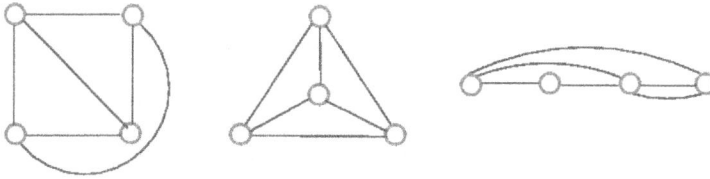

1.2.11 Regular Graph

A graph is said to be regular if all vertices of graph have same degree. If degree of each vertex is r, then G is regular of degree r.

Graph with degree r = 0

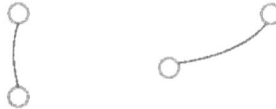

Graphs with degrees r = 2

Graphs with degree r = 2

Graphs with degrees r = 3

Graphs with degrees r = 4

1.2.12 Isomorphic Graphs

Graphs are said to be isomorphic, when second graph can be obtained from first graph by relabeling the vertices. If there is one-to-one correspondence among vertices of first graph with those of second graph such that number of edges joining any pair of vertices in second graph will be equal to the number of edges joining the corresponding pair of vertices in second graph. In an example where two unlabeled graphs are shown

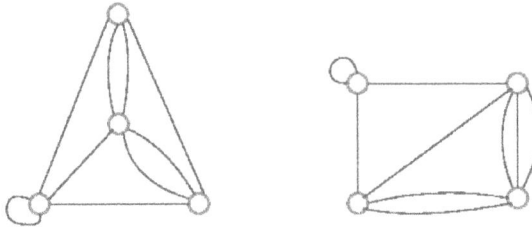

to be isomorphic when their labels can be attached to their vertices so that they become the same graph.

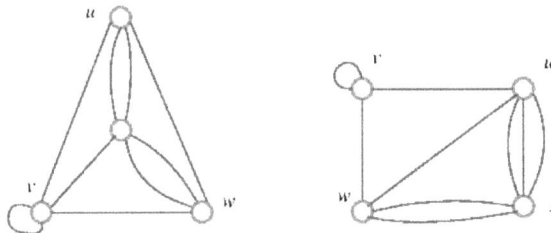

Figure 1.26: Isomorphic Graphs.

1.3 Network Graph Components

A graph is a set of vertices and a collection of edges that each connects a pair of vertices. We use the names 0 through V-1 for the vertices in a V-vertex graph (Figure 1.27).

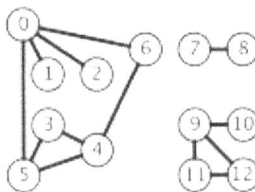

Figure 1.27: Vertex and Edges in Graph.

Nodes

The nodes in a graph shows object and the lines shows the relationships. The line between the two points represents node (Figure 1.28).

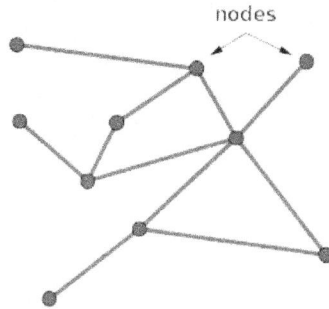

Figure 1.28: Nodes in the Graph.

Vertices

Vertex is a fundamental unit of which graphs are formed. In a undirected graph, there are set of vertices and set of edges, while in a directed graph, there are set of vertices and set of arcs. A vertex is shown by a circle with a label where an edge is shown by a line or arrow extending from one vertex to another (Figure 1.29).

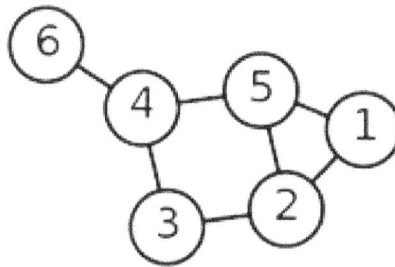

Figure 1.29: Graph with 6 Vertices and 7 Edges.

Vertices can be:

Isolated vertex:
It is a vertex with degree zero, it means that a vertex which is not an endpoint of any edge

Leaf vertex:
It is a pendant vertex. having a degree one. In directed graph, one can distinguish outdegree from indegree.

Source vertex:
It is a vertex with indegree zero

Sink vertex:
It is a vertex with outdegree zero

Simplicial vertex:
It is a vetrex whose neighbours form a clique where every two neighbours are adjacent.

Universal vertex:
It is a vertex which is adjacent to every other vertex in a graph (Figure 1.30).

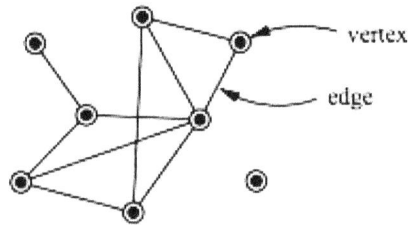

Figure 1.30: Graph with 8 Vertices and 9 Edges (1 isolated).

Cut Point

A cutpoint is a vertex whose removal from the graph increases the number of components. That is, it makes some points unreachable from some others. It disconnects the graph. In a graph shown, a vertex D is disconnected and hence the graph gets disconnected to two disjoint graphs forming E as another cut point (Figure 1.31).

Figure 1.31: Cut Point of Graph.

Cut Set

A cutset is a collection of points whose removal increases the number of components in a graph. A minimum weight cutset consists of the smallest set of points that must be removed to disconnect a graph. The number of points in a minimum weight cutset is called the point connectivity of a graph. If a graph has a cutpoint, the connectivity of the graph is 1. The minimum number of points separating two nonadjacent points s and t is also the maximum number of point-disjoint paths between s and t.

Bridge

A bridge is an edge whose removal from a graph increases the number of components (disconnects the graph). An edge cutset is a collection of edges whose removal disconnects a graph. A local bridge of degree k is an edge whose removal causes the distance between the endpoints of the edge to be at least k. The edge-connectivity of a graph is the minimum number of lines whose removal would disconnect the graph. The minimum number of edges separating two nonadjacent points s and t is also the maximum number of edge-disjoint paths between s and t (Figure 1.32).

Figure 1.32: Bridge in Network Graph.

In this, bridge edge {D,E}is removed where {E,F} and {E,G} also serves as bridge edges. It is noted that every bridge edge has a cut point at one end. A network can be shown as a graph whose vertices passes information while edges shows communications links over which data travels.

Loop

Loop is an edge with endpoints equal where an edge joins a vertex to itself. A graph has multiple edges when two or more edges in a graph joins the same pair of vertices (Figure 1.33).

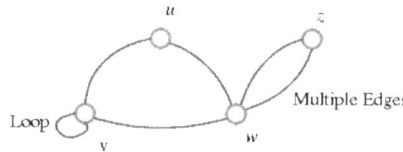

Figure 1.33: Loop in a Graph.

Trail and Path

If all edges of a walk are different, then walk is known as a trail. Further when all vertices are difficult, then trail is known as path (Figure 1.34).

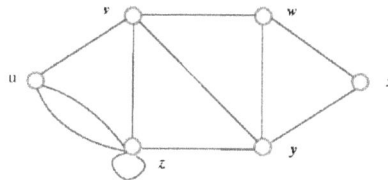

Figure 1.34: Trail and Path in a Graph.

In the above graph:

➤ Walk vzzywxy is a trail as vertices y and z occurs twice.

➤ Walk vwxyz is a path as walk has no repeated vertices.

1.4 Representation of Graphs

1.4.1 Adjacency Matrix

Adjacency matrix is a **V x V** binary matrix **A** (a binary matrix is a matrix in which the cells can have only one of two possible values - either a 0 or 1). Element $A_{i,j}$ is

1 if there is an edge from vertex i to vertex j else $A_{i,j}$ is 0. The adjacency matrix can also be modified for the weighted graph in which instead of storing 0 or 1 in Ai,j we will store the weight or cost of the edge from vertex i to vertex j.

In an undirected graph, if $A_{i,j}$ = 1 then $A_{j,i}$ = 1. In a directed graph, if $A_{i,j}$ = 1 then $A_{j,i}$ may or may not be 1. Adjacency matrix providers **constant time access** **(O(1))** to tell if there is an edge between two nodes. Space complexity of adjacency matrix is **O(V²)** (Figure 1.35).

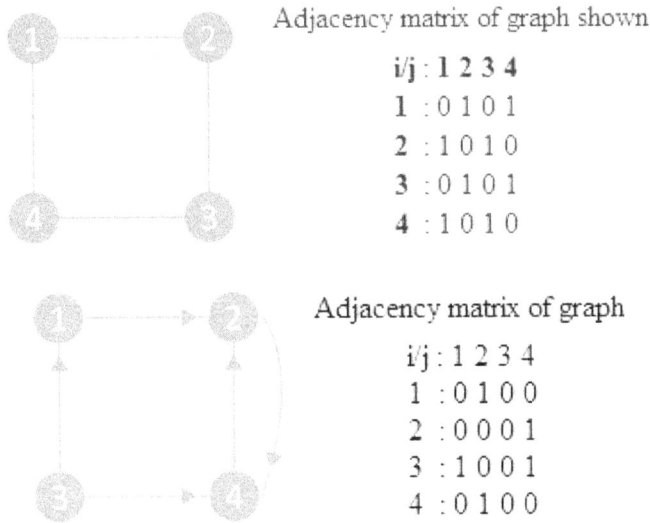

Adjacency matrix of graph shown

i/j : **1 2 3 4**
1 : 0 1 0 1
2 : 1 0 1 0
3 : 0 1 0 1
4 : 1 0 1 0

Adjacency matrix of graph

i/j : 1 2 3 4
1 : 0 1 0 0
2 : 0 0 0 1
3 : 1 0 0 1
4 : 0 1 0 0

Figure 1.35: Graphs with Adjacent Matrix.

In an above directed graph using Adjacency matrix, edges of the graph will be shown. Adjacency matrix is very convenient to work with. Add (remove) an edge can be done in O(1) time, the same time is required to check, if there is an edge between two vertices.

By running the program using input file details as shown, it is observed that:

Input File:
4
5
1 2
2 4
3 1
3 4
4 2
Output:
There is an edge between 3 and 4
There is no edge between 2 and 3

1.4.2 Adjacency List

The other way to represent a graph is an adjacency list. Adjacency list is an array A of separate lists. Such type of graph representation is one of the alternatives to adjacency matrix where less amount of memory will able outperform adjacency matrix. For every vertex adjacency list stores a list of vertices, which are adjacent to current one.

Consider an example

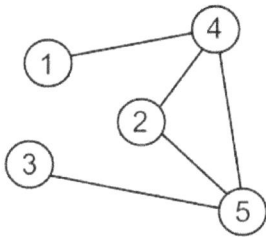

| Graph | Adjacency list |

Graphs with Adjacent Matrix

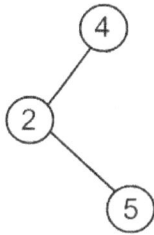

Vertices, adjacent to {2}   Row in the adjacency list

Graphs with Adjacent Matrix.

Adjacent list allows us to store graph in more compact form, than adjacency matrix, but the difference decreasing as a graph becomes denser. Next advantage is that adjacent list allows to get the list of adjacent vertices in O(1) time, which is a big advantage for some algorithms.

1.5 Network Model

The network model is graphical in that it is presented as a collection of the nodes and arcs drawn in the figure. The nodes represent the cities of this problem, and we name them with the shortened names of the supply and demand cities. Arcs are the directed line segments of the figure. The nodes at its ends identify an arc.

Flow is associated with the network, entering and leaving at the nodes and passing through the arcs. Flows entering or leaving a node from external sources are called external flows and are shown adjacent to the nodes in the square brackets. A positive external flow is a supply, flow that enters the network, and a negative external flow is a demand, flow that leaves the network. Flow is conserved at each node, implying that the total flow entering a node, either from arcs or external supplies, must equal the total flow leaving the node, either to arcs or to external demands. The arc flows are the decision variables for the network flow programming model.

Flow is limited in an arc by the lower and upper bounds on flow. In this example we specify 0 as the lower bound for all arcs and 200 as the upper bound. Sometimes the term capacity refers to the upper bound on flow. Within the restrictions imposed by conservation of flow for each node and the bounds on flow for each arc, there are usually many feasible flows (a flow is an assignment of an arc flow to each arc). The problem is to find a feasible flow, if one exists, and an optimal flow from the set of feasible flows.

The criterion for optimality is cost. Associated with each arc is a cost per unit of flow (the number in the parenthesis). If a flow x passes through the arc with unit cost c, a cost cx is incurred. The total cost for the network is the sum of the arc costs, and the goal of optimization is to find the feasible flow that minimizes this measure.

We call the model of this section a pure network flow model because the flow entering an arc at its origin node is equal to the flow leaving the arc at its terminal node. This is contrasted later to the generalized network flow model that does not have this limitation. The pure model has the important feature of integral optimum solutions. Whenever all node external flows and all arc upper and lower bounds are integer, the flow solution to the pure model is also integer. As we will see, this has important ramifications.

1.5.1 Linear Programming Model

Every network flow model has a linear programming model, that is a model with algebraic linear expressions describing the objective function and constraints (Figure 1.36).

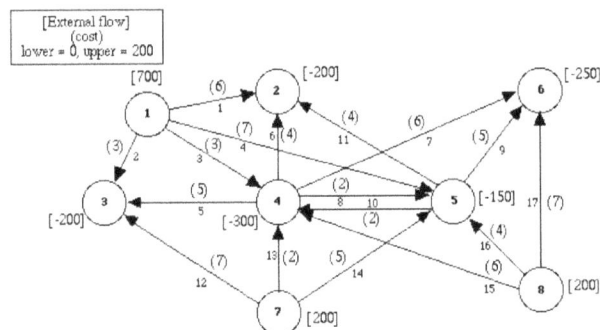

Figure 1.36: Representation for Linear Programming Model.

For construction of the model, it is convenient to number the nodes and arcs for reference

The linear programming model is an algebraic description of the objective to be minimized and the constraints to be satisfied by the variables. The variables are the flows in each arc designated by x1 through x17. The network flow problem is to minimize total cost while satisfying conservation of flow at each node. The variables must also satisfy the simple upper and lower bounds on arc flow.

The objective is to minimize total cost

$$z = 6x_1 + 3x_2 + 3x_3 + 7x_4 \ldots + 7x_{17}$$

The main constraints require conservation of flow at each node:

Node 1: $x_1 + x_2 + x_3 + x_4 = 700$

Node 2: $-x_1 - x_6 - x_{11} = -200$

Node 3: $-x_2 - x_3 - x_{12} = -200$

Node 4: $x_3 + x_6 + x_7 + x_3 - x_3 - x_{10} - x_{13} = -300$

$\vdots$ $\vdots$

Node 8 $x_{13} + x_{16} + x_{17} = 200$

Simple lower and upper bounds: $0 \le x_j \le 200$ for $j = 1 \ldots 17$.

Generalized Network Model

We introduce an additional parameter called the arc gain to handle losses or gains that occur along an arc. The arc gain is an arc parameter that multiplies the flow at the beginning of the arc to obtain the flow at the end of the arc. Figure 1.37 shows the effect of the gain on flow.

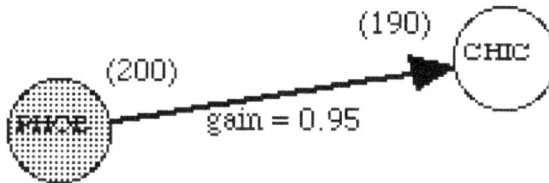

Figure 1.37: Modeling Losses with a Gain Factor.

The solution with loses in next figure is significantly different than those previously presented. The flows are no longer integer, the extra demand at Chicago is no longer satisfied, Austin does not produce to its full capacity, more arcs are used to provide materials lost during shipping and the profit is considerably reduced.

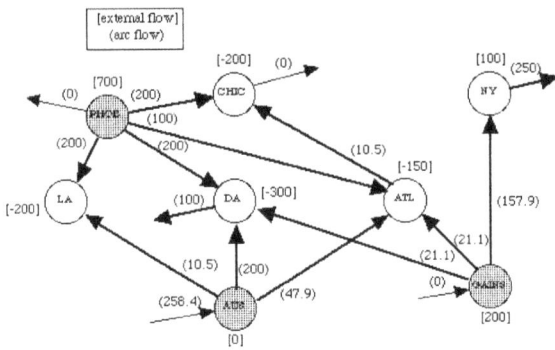

Figure 1.38

Gains are very for useful modeling. When all arc gains are 1, the model is a pure network flow model. When some gains are other than 1, the model is a generalized network flow model. Integer solutions cannot be guaranteed for the generalized model.

2 | Multi-terminal Multipath Flows

2.1 Multi-terminal Maximal Flows

The problem of finding the maximum flows between each pair of nodes in a subset of k nodes of a node and arc-capacitated undirected network arises in the study of communication and transportation networks. It is shown that in arc-capacitated case, it suffices to solve only k-1 of the k(k- 1)/2 required maximum-flow problems. Apparently the node-capacitated case cannot be reduced to arc-capacitated case. Indeed, the reduction of an undirected node capacitated network to an arc-capacitated one leads to a network that is not symmetric, *i.e.*, the capacity of an arc joining two nodes depends upon the direction in which the arc is traversed.

By contrast, symmetry is not lost in reducing the arc to node-capacitated problem. this reduction entails replacing each capacitated arc by a pair of un-capacitated arcs incident to a node whose capacity equals the capacity of the original arc, hence the networks with node capacities in the sequel are only considered.

Example 2.1

In a directed network shown with source O and sink T, what will be the maximum flow from O to T in network if path O→B→D→C→T is selected as first augment path.

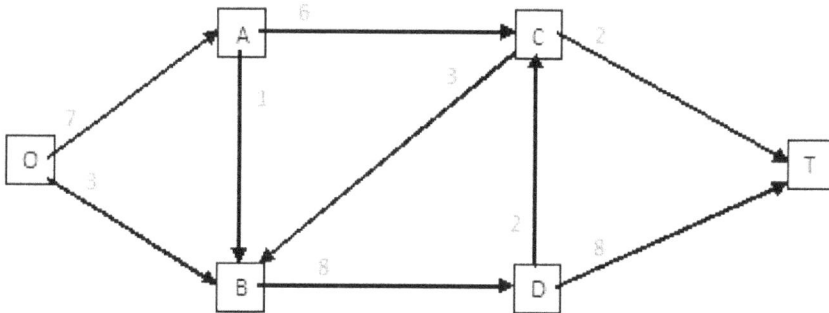

Maximum flow from O to T.

Solution

Considering first residual network:

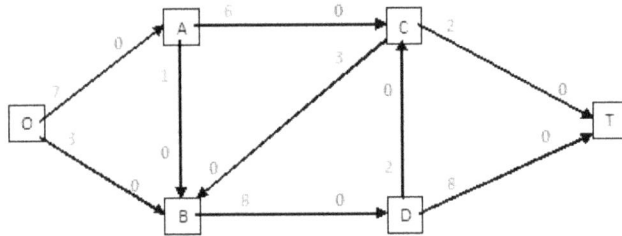

Residual Network.

As required, O→B→D→C→T is selected as first augmenting path as this path creates necessity for reversing some flow in later iterations. Its residual capacity is min{3,8,2,2}=2. After sending 2 units of flow through the path, the resulting residual network is:

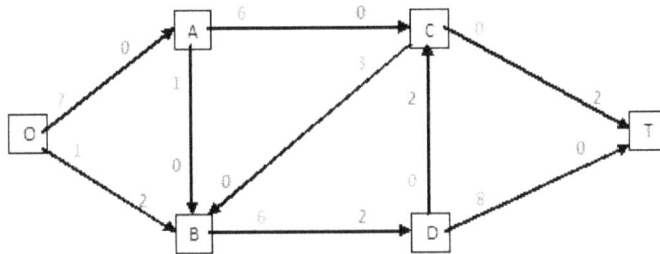

Selecting First Augmenting Path.

Now select path O→A→C→B→D→T as next augmenting path. In this, the residual capacity of the path is min{7,6,3,6,8}=3, so after sending 3 units of flow through the path, the new residual network will be obtained as:

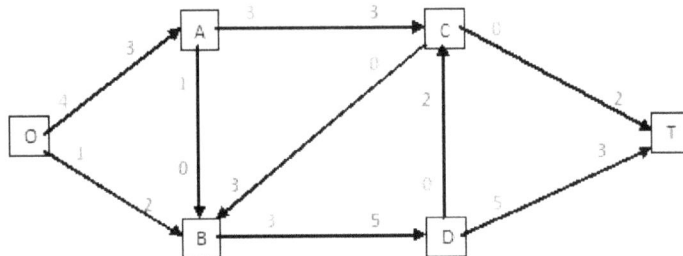

Selecting Second Augmenting Path.

Further select path O→B→D→T as third augmenting path in which the residual capacity is min{1,3,5}=1. Hence after sending 1 units of flow through the path, new residual network will result as shown:

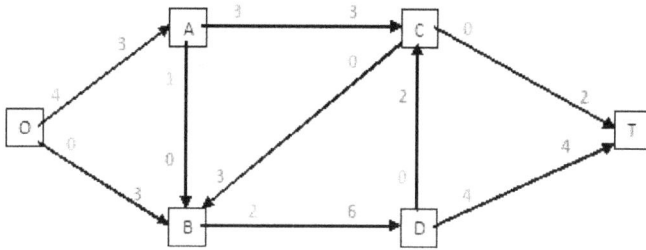

Selecting Third Augmenting Path.

Now select fourth augmenting path as O→A→B→D→T where the residual capacity is min{4,1,2,4}=1. After sending 1 units of flow through the path, the new residual network is:

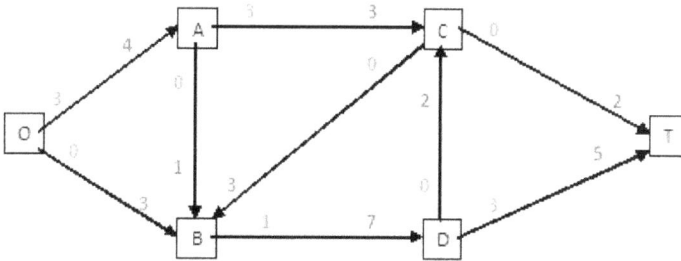

Selecting Fourth Augmenting Path.

Finally select O→A→C→D→T as the augmenting path where residual capacity is min{3,3,2,3}=2. Here, the flow is reverse on arc D→C, hence after sending 2 units of flow through the path, the new residual network will result as:

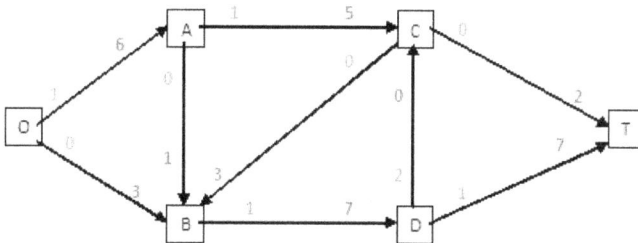

Selecting Fifth Augmenting Path.

It is observed that there will be no more augmenting paths from O to T that are left, hence the current residual network is optimal. The numbers 1, 3, 6, 5, 0, 7 and 2 on the arcs show the optimal flow values. Maximum flow value is 7+2 = 9.

2.1.1 Maximum Flow Problem

Maximum flow problem is a structure on a network where the arc capacities or upper bounds are relevant parameters. The problem here is to find the maximum flow possible from some given source node to a given sink node. A network model as shown in figure below with arc costs zero, where cost on arc leaving the sink is set to -1. As the goal of optimization is to minimize the cost, so the maximum flow possible is delivered to the sink node (Figure 2.1).

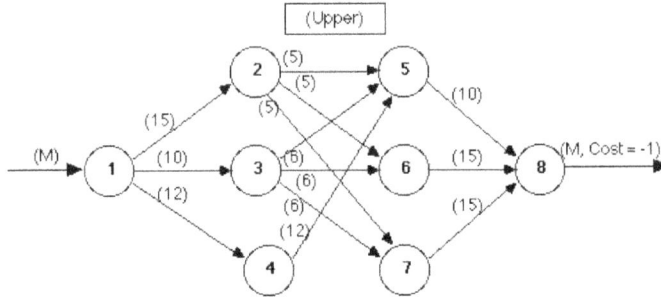

Figure 2.1: Network Model for the Maximum Flow Problem.

The solution to the example is shown with Figure 2.2 where maximum flow from node 1 to node 8 is 30. The heavy arcs on the figure are called the minimal cut.

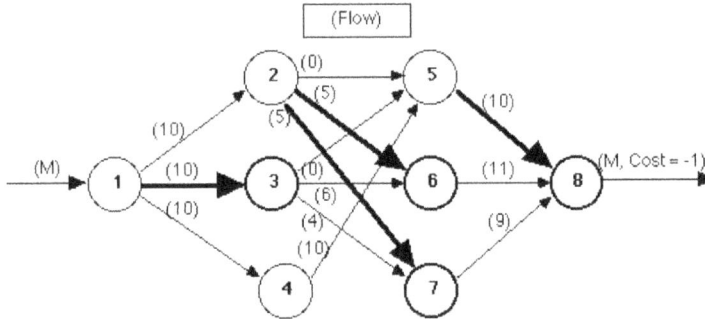

Figure 2.2: Describing Maximum Flow.

These arcs are bottlenecks that restrict the maximum flow. It is the fact that the sum of the capacities of the arcs on the minimal cut will be equal to the maximum flow which serves as a famous theorem of network theory called as max flow min cut theorem. The arcs on the minimum cut can be identified using sensitivity analysis.

2.2 Minimum Cut Capacities

It is observed that every day, a parcel service transports maximum number of parcels from city s to city t. Each transport has to pass the intermediate cities a, b, c, or d, as shown in figure. Each individual transport route say e=(x,y) from city x to city y has only certain limited capacity where only limited number cap(e) of parcels can be transported from city x to city y through e per day, as only certain number of trucks are available on this route (Figure 2.3).

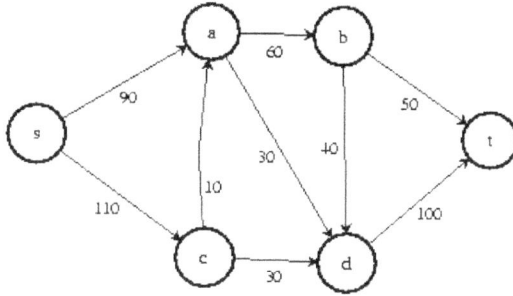

Figure 2.3: Edges Labeled with Capacity and Source of Network.

Here the number of parcels that can be forwarded from city x per day cannot be larger than the number of parcels that are arriving there. In intermediate cities, all arriving parcels must be forwarded each day without any stacking parcels. In forwarding process, capacity of a transport route will never be exceeded though only in city t, sink parcels may be opened and used up as they vanishes, so parcels may be created and packed only in city s, source as they come into existence there.

The problem of parcel service is that it cannot be modified as per the capacities of the transport routes but only defines the forwarding of parcels, as how many parcels shall flow per day over each transport route without capacities being exceeded. Here the flow needs to be defined such that maximum possible number of parcels arrive in city t per day. Also, to generate maximum flow from s to t, it is not essential that parcels are transported from s to t as fast as possible but it is required that as many parcels as possible vanish in sink per day.

In a directed graph G, edges e carry a non-negative capacity information cap(e) is a network. Let s, the source, and t, the sink, be two different nodes of G. A network flow from s to t in such a network is a mapping f that maps each edge e a value f(e) such that the following conditions are satisfied:

Capacity Constraint

0 <= f(e) <= cap(e), that is, no flow along an edge is negative or exceeds the feasible capacity.

Flow Conservation

For each node except s and t the following holds: The sum of f(e) over all edges e incident to v is equal to the sum of the f(g) over all edges g leaving v. In physics it is known as Kirchhoff's Law which explains that the sum of currents entering any node will be equal to the sum of the currents leaving that node.

The value of a flow f is arrives at t, but does not vanish there will be the sum of f(e) over all edges e incident to t minus the sum of f(g) over all edges g leaving t. Due to the flow conservation, value of flow will be equal to the sum of flows leaving s minus the sum of the flows entering s (Figure 2.4).

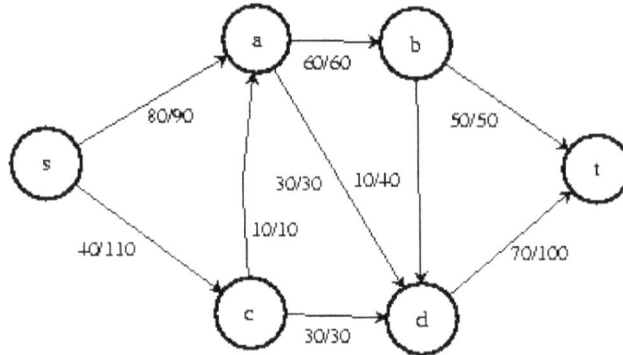

Figure 2.4: Maximum Flow for the Network.

A maximum flow from s to t is a flow with a maximum value. It can be shown that there is always a flow with a maximum value in which the sum of the flows leaving t and the sum of the flows entering s both are 0.

2.2.1 Minimum Cuts

A cut of a network divides it into two non-empty parts which shows a partition (S,T) of node set V into two non-empty subsets S and T. A cut edge is an edge that is crossed by the cut, whose source belongs to S and whose target belongs to T or vice-versa. An s-t cut is a cut (S,T) that separates the source s from the sink t in sets S and T. The value of a cut is the sum of all weights (capacities) of all edges that cross the cut from S to T, that is, whose source belongs to S and whose target belongs to T. A minimum cut is a cut of minimum value (Figure 2.5).

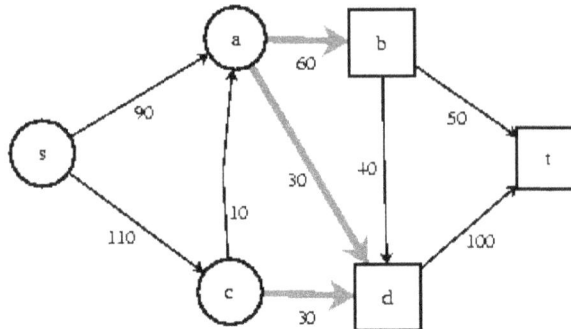

Figure 2.5: A Minimum Cut.

A well-known theorem from graph theory states that the maximum flow is always equal to the minimum s-t-cut. Furthermore, the value of every flow is less or equal to the value of every s-t-cut. Therefore the cut from above figure is also a minimum s-t-cut. The value of this cut is 120, and since we see a flow also with value 120 in figure shown below, there can be neither a smaller s-t-cut, nor a larger one; so this proves not only the minimality of s-t-cut from Figure 2.5, but also the maximality of flow from Figure 2.6.

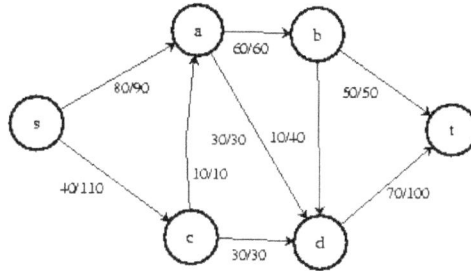

Figure 2.6: Maximality of Flow.

Example 2.2

Find the minimum cut capacity of directed network shown with source O and sink T, if path O→B→D→C→T is selected as first augment path.

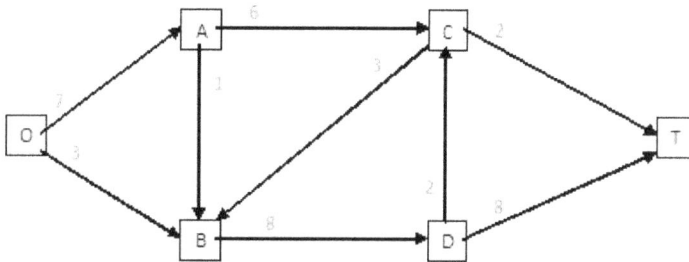

Maximum Flow from O to T.

Solution

In a directed network as shown where source O and sink T, for finding minimum cut capacity, the graph should be plotted to find the maximum flow from O to T in network initially by considering the path O→B→D→C→T as its first augment path, so first residual network will have:

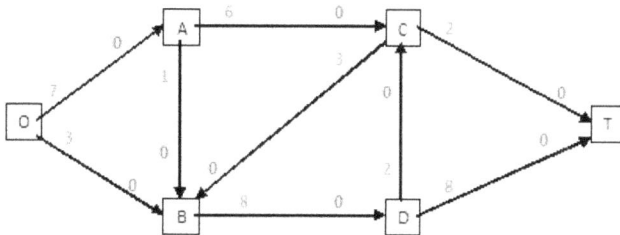

Residual Network.

As required, O→B→D→C→T is selected as first augmenting path as this path creates necessity for reversing some flow in later iterations. Its residual capacity is min{3,8,2,2}=2. After sending 2 units of flow through the path, the resulting residual network is:

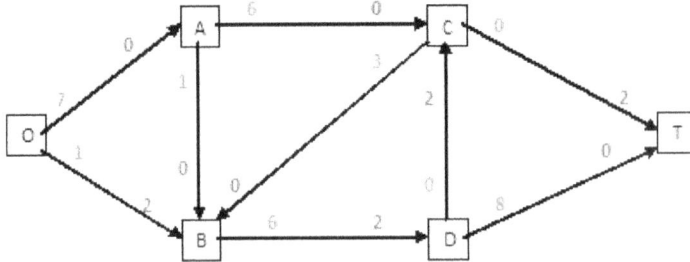

Selecting First Augmenting Path.

Now select path O→A→C→B→D→T as next augmenting path. In this, the residual capacity of the path is min {7,6,3,6,8}=3, so after sending 3 units of flow through the path, the new residual network will be obtained as:

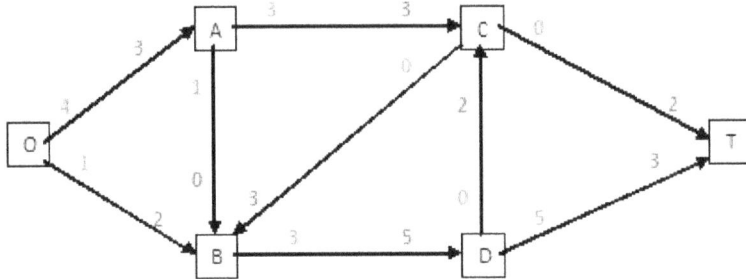

Selecting Second Augmenting Path.

Further select path O→B→D→T as third augmenting path in which the residual capacity is min{1,3,5}=1. Hence after sending 1 units of flow through the path, new residual network will result as shown:

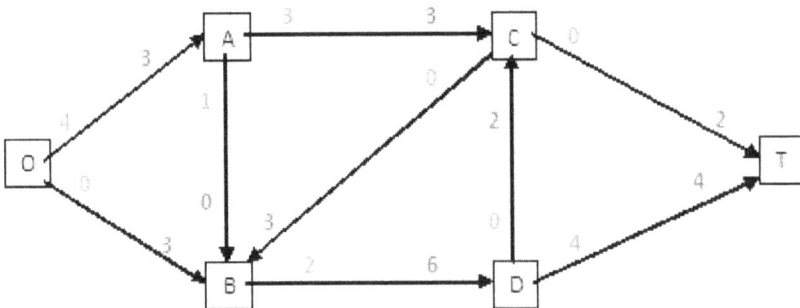

Selecting Third Augmenting Path.

Now select fourth augmenting path as O→A→B→D→T where the residual capacity is min{4,1,2,4}=1. After sending 1 units of flow through the path, the new residual network is:

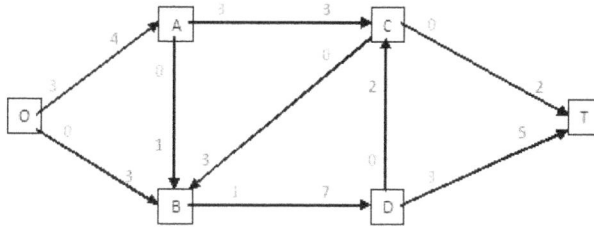

Selecting Fourth Augmenting Path.

Finally select O→A→C→D→T as the augmenting path where residual capacity is min{3,3,2,3}=2. Here, the flow is reverse on arc D→C, hence after sending 2 units of flow through the path, the new residual network will result as:

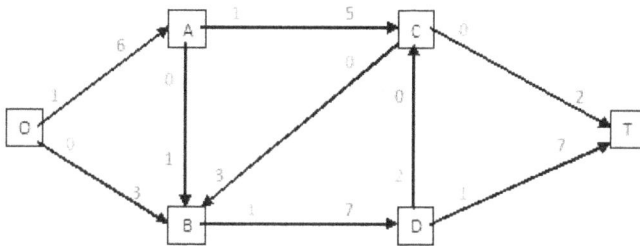

Selecting Fifth Augmenting Path.

It is observed that there will be no more augmenting paths from O to T that are left, hence the current residual network is optimal. The numbers 1, 3, 6, 5, 0, 7 and 2 on the arcs show the optimal flow values. Maximum flow value is 7+2 = 9.

From the above network graph where maximum flow value is 9, the nodes which can be reached from source by augmenting paths will be A and C. Hence the O-side of minimum cut is {O, A, C}, so minimum cut will result as:

MinCut = {O→B, A→B, C→B, C→T}

Here it is observed that arc D→C is not in cut as it goes from T-side to O-side. Here the capacity of minimum cut is 3+1+3+2 = 9 which is equal to maximum flow value, hence the minimum cut shown in the network will be:

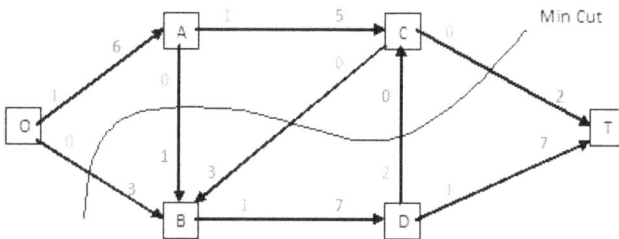

Minimum Cut Graph.

2.3 Multi-terminal Shortest Paths

2.3.1 Shortest Paths

Shortest path between two nodes in a graph is a path with minimum number of edges. If the graph is weighted, it is a path with the minimum sum of edge weights. The length of shortest path is called geodesic distance or shortest distance. Shortest paths are not necessarily unique, but the geodesic distance is well-defined since all geodesic paths have the same length.

The large scale property of networks results as average geodesic distance between pairs of nodes in network. An important network phenomena is small-world effect where in many real networks, typical geodesic distance is short when compared with the number of nodes of the network.

It is observed that small world effect has a substantial implication for networked systems. In case of spreading of rumor over a social network, it is clearly noticed that the rumor will reach to the people much faster when it is only six steps from any person to any other instead of hundred. Similarly, the speed with which one gets a response from another computer on the Internet depends on how many hops data packets have to make as they traverse the network.

In fact, once one looks deeply into the mathematics of networks, one discovers that the small-world effect is not so surprising after all. Mathematical models of networks suggest that path lengths in networks typically scale with number of network nodes should tends to remain small even for large networks. The idea behind short distances is that the number of nodes one encounters in a breath-first visit of a graph starting from a source node typically increases exponentially with the distance to the source node, hence the distance increases logarithmically in the number of visited nodes.

Analysing the diameter of network, which serves as largest geodesic distance on graph where it is normally very small with calculations that suggests that network models should be scaled logarithmically with number of nodes as the average geodesic distance. The diameter is, however, a less useful measure for real networks than the average distance, since it really only measure the distance between one pair of vertices at the extreme end of the distribution of distances. Moreover, the diameter is sensible to small changes to the set of nodes, which makes it a poor indicator of the behavior of the network as a whole. By analyzing average geodesic distance and diameter of random graph using an algorithm, it is seen that:

```
n = 100
p = 1.5/n
g = erdos.renyi.game(n, p)
# average geodesic distance
average.path.length(g)
[1] 5.773585
# diameter
```

```
d = get.diameter(g)
length(d) - 1
[1] 14
E(g)$color = "grey"
E(g)$width = 1
E(g, path=d)$color = "red"
E(g, path=d)$width = 2
V(g)$color = "SkyBlue2"
V(g)[d]$color = "red"
coords = layout.fruchterman.reingold(g)
plot(g, layout=coords, vertex.label = NA, vertex.size=3)
```

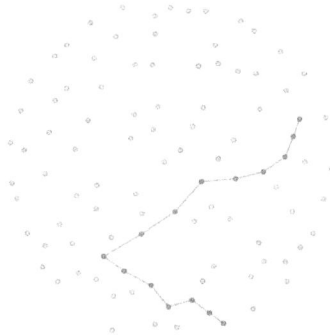

Figure 2.7: Geodesic Distance and Diameter of Random Graph.

It is observed that the idea of the shortest distance can also be analysed by using bar graph where share of geodesics be plotted in terms of length having a typically shorter lengths as compared to number of nodes where 19% of geodesics have length 5, 33% have length 6, and 26% have length 7. Here, the average geodesic distance is 6.41 and distances normally distribute around the peak. The largest distance is also small as 23 having 8 different geodesics with such length (Figure 2.8).

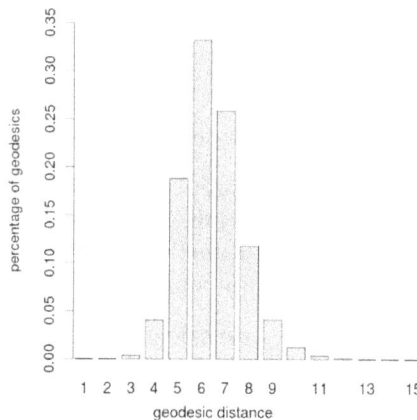

Figure 2.8: Bar Graph of Geodesic Distance.

The fact that distances normally distribute is interesting because it means that the average distance of 6 links represents a typical value of all distances in the network. Furthermore, since the distribution of distances drops off rapidly around the mean, the time-consuming computation of the exact average distance can be approximated by computing the average distance on a relatively small random sample of node pairs.

It has been proposed to use paths on social networks as referral chains to establish contacts with domain experts. In general, we might use the intermediate people in referral chains in order to smoothly get in contact with the target person.

From the above bar graph, it is observed that the average weighted distance on the network will be 3.15, and average length of the weighted geodesics is 11.27. Hence, light paths are much longer, almost twice longer, than the typical geodesic path, which is about 6 edges long. Furthermore, the average geodesic distance is 6.43, and the average weight of the geodesics is 5.07. With this, the short paths are significantly heavier than the typical weighted geodesic path, which weights about 3, so the weighted shortest paths and unweighted shortest paths are different referral chains.

2.3.2 Shortest Paths between Pairs of Nodes in Network

Shortest paths among nodes are the main elements of network research which serves as key component of number of measures. It describes the average distance among places and people and are used as underlying metric in number of measures between number of shortest paths and among others that passes through a node (Figure 2.9).

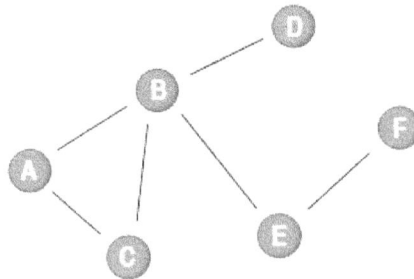

Figure 2.9: One-Mode Binary Network.

The shortest distance among nodes in binary one-mode networks is able to calculate the number of ties along the shortest path. If two nodes are directly connected, then distance=1 while if they are not directly connected, but are joined through intermediaries, then in such, lowest number of intermediary nodes will be +1.

To exemplify how shortest distances are calculated, the distance between node A and F in the sample network as shown above is 3 that shows minimum number of ties between them. The distances between all node pairs are shown in matrix. The average shortest distance is calculated by taking the mean of the matrix excluding the diagonal. In this cases, it is 1.8.

In two-mode network as shown in Figure 2.10, shortest paths have been identified in a number of ways. While the methods vary which lead to substantial similar outcome where distance among two nodes results equal to the number of ties among them.

	[A]	[B]	[C]	[D]	[E]	[F]
[A]	NA	1	1	2	2	3
[B]	1	NA	1	1	1	2
[C]	1	1	NA	2	2	3
[D]	2	1	2	NA	2	3
[E]	2	1	2	2	NA	1
[F]	3	2	3	3	1	NA

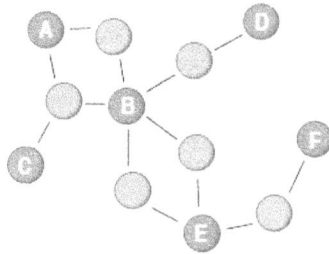

Figure 2.10: Network with Two Types of Nodes.

As observed, to represent two-mode networks in (n+p)x(n+p)-matrices where n and p are the numbers of primary and secondary nodes, respectively, instead of rectangular two-mode matrices (n x p), such matrix type is extremely inefficient as it is at least four times larger than the rectangular one without holding any additional information. Nevertheless, it serves as key feature in case of a square. As such, one-mode metrics like one-mode shortest path algorithms be applied to it in which two-mode sample network can be shown in matrix as below:

	[A]	[B]	[C]	[D]	[E]	[F]	[1]	[2]	[3]	[4]	[5]	[6]
[A]	0	0	0	0	0	0	1	1	0	0	0	0
[B]	0	0	0	0	0	0	1	1	1	1	1	0
[C]	0	0	0	0	0	0	0	1	0	0	0	0
[D]	0	0	0	0	0	0	0	0	1	0	0	0
[E]	0	0	0	0	0	0	0	0	0	1	1	1
[F]	0	0	0	0	0	0	0	0	0	0	0	1
[1]	1	1	0	0	0	0	0	0	0	0	0	0
[2]	1	1	1	0	0	0	0	0	0	0	0	0
[3]	0	1	0	1	0	0	0	0	0	0	0	0
[4]	0	1	0	0	1	0	0	0	0	0	0	0
[5]	0	1	0	0	1	0	0	0	0	0	0	0
[6]	0	0	0	0	1	1	0	0	0	0	0	0

A second method is to project two-mode networks to one-mode networks (n x n-matrix), and then applying the standard one-mode shortest path algorithm. Projection works connecting primary nodes that a connected to common nodes in the two-mode structure. To exemplify this structure, the matrix below is the binary projection of the two-mode network above.

	[A]	[B]	[C]	[D]	[E]	[F]
[A]	0	1	1	0	0	0
[B]	1	0	1	1	1	0
[C]	1	1	0	0	0	0
[D]	0	1	0	0	0	0
[E]	0	1	0	0	0	1
[F]	0	0	0	0	1	0

Using both the above procedures produces a similar outcome. The only difference is that all distances found using the first method are twice the distances found using the second method. For example, the distances found from node B to node F are 4 and 2 using the two methods, respectively.

There are a number of issues with these two methods:

> Two nodes connected through two common nodes (B and E) are assumed to have an identical connection as two connected through only one common node (B and D).

> All non-projected nodes are considered equal. This implies that nodes connecting many nodes are equal to the ones connecting only two nodes.

> The secondary nodes connecting few primary nodes gave those nodes a higher level of interaction.

There is no established way of calculating shortest paths in weighted two-mode network.

Building on the second method, projection, it is possible to deal with some of these limitations. By projecting two-mode networks (both binary and weighted) onto weighted one-mode networks, and then apply the weighted shortest path algorithm, this procedure has the possibility of capturing more information than simply saying that weights do not matter. It is important to use an appropriate projecting method as this will influence the outcome of the measure. To exemplify this method, the distances in the two-mode sample network using the sum-projection method would be:

	[A]	[B]	[C]	[D]	[E]	[F]
[A]	NA	0.67	1.33	2.00	1.33	2.67
[B]	0.67	NA	1.33	1.33	0.67	2.00
[C]	1.33	1.33	NA	2.67	2.00	3.33
[D]	2.00	1.33	2.67	NA	2.00	3.33

[E]	1.33	0.67	2.00	2.00	NA	1.33
[F]	2.67	2.00	3.33	3.33	1.33	NA

From the above matrix, the distance between B and E (0.67) is half of the distance from B to D (1.67). This is because they have two common nodes instead of only one, and thus, have a stronger connection.

2.3.3 Shortest Paths Problem

This problem uses a general network structure where only the arc cost is relevant. A typical case is shown in figure below where length of a path is the sum of arc costs along the path. In this, the problem finds the shortest path from some specified node to some other node or perhaps to all other nodes. The latter problem is called the shortest path tree problem as it is the collection of all shortest paths from a specified node forms a graph structure called a tree. Since it is not much more difficult to find the paths to all nodes than it is to find the path to one node, the shortest path tree problem is usually solved (Figure 2.11).

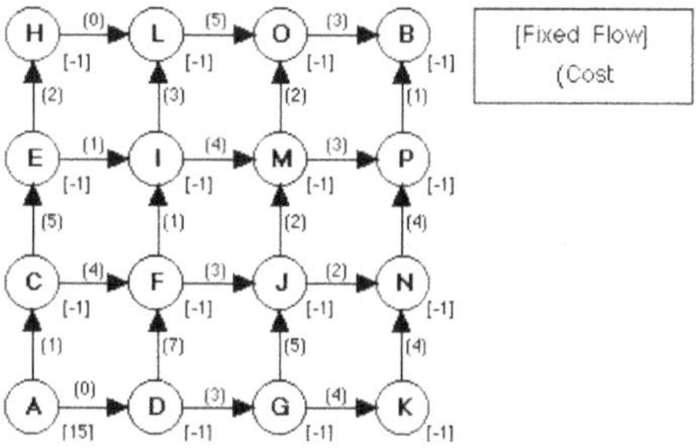

Figure 2.11: Network Model for Shortest Path Problem.

The network flow equivalent to the shortest path tree problem is formed by equating arc distance to arc cost, assigning a fixed external flow of m - 1 (m is the number of nodes) to the source node, and assigning fixed external flows of -1 to every other node. When solving this flow problem, the computer will assign flow from the source to each node by the least cost path, since there are no bounds on arc flows. The shortest path tree will consist of those arcs with nonzero flow in the optimum solution as shown in Figure 2.12.

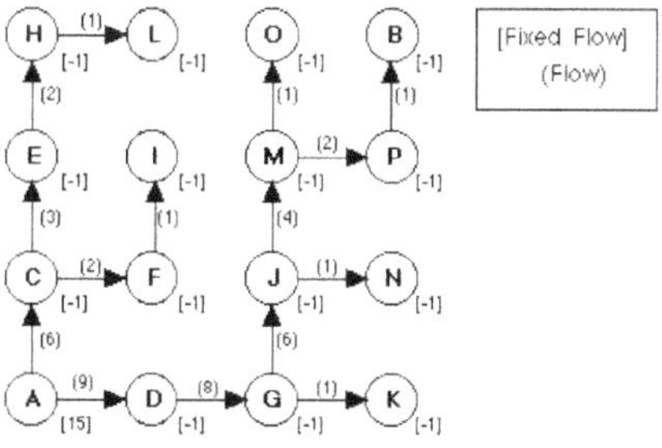

Figure 2.12: Solution of Shortest Path Problem.

2.3.3 Algorithm for Shortest Path difference in Network

Shortest path algorithms are a family of algorithms designed to solve the shortest path problem. The shortest path problem is something most people have some intuitive familiarity with: given two points, A and B, what is the shortest path between them? In computer science, however, the shortest path problem can take different forms and so different algorithms are needed to be able to solve them all.

For simplicity and generality, shortest path algorithms typically operate on some input graph, G. This graph is made up of a set of vertices, V and edges E that connect them. If the edges have weights, the graph is called a weighted graph. Sometimes these edges are bidirectional and the graph is called undirected. Sometimes there can be even be cycles in the graph. Each of these subtle differences are what makes one algorithm work better than another for certain graph type. An example of a graph is shown in Figure 2.13.

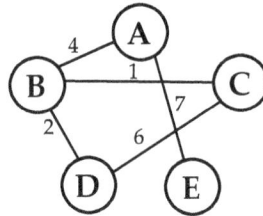

Figure 2.13: Example of Shortest Path Algorithm a Graph.

There are also different types of shortest path algorithms. Maybe you need to find the shortest path between point A and B, but maybe you need to shortest path between point A and all other points in the graph.

Shortest path algorithms have many applications. As noted earlier, mapping software like Google or Apple maps makes use of shortest path algorithms. They are also important for road network, operations, and logistics research. Shortest path algorithms are also very important for computer networks, like the Internet.

Types of Shortest Path Algorithms

There are two main types of shortest path algorithms, single-source and all-pairs. Both types have algorithms that perform best in their own way. All-pairs algorithms take longer to run because of the added complexity. All shortest path algorithms return values that can be used to find the shortest path, even if those return values vary in type or form from algorithm to algorithm.

Single-source

Single-source shortest path algorithms operate under the following principle:

Given a graph G, with vertices V, edges W with weight function $w(u,v) = w_{u,v}$ and a single source vertex, s return the shortest paths from s to all other vertices in V.

If the goal of the algorithm is to find the shortest path between only two given vertices, and, then the algorithm can simply be stopped when that shortest path is found. Because there is no way to decide which vertices to "finish" first, all algorithms that solve for the shortest path between two given vertices have the same worst-case asymptotic complexity as single-source shortest path algorithms.

This paradigm also works for the single-destination shortest path problem. By reversing all of the edges in a graph, the single-destination problem can be reduced to the single-source problem. So, given a destination vertex,, this algorithm will find the shortest paths starting at all other vertices and ending at .

All-pairs

All-pairs shortest path algorithms follow this definition:

Given a graph G, with vertices V, edges E with weight function $w(u,v) = w_{u,v}$ return the shortest path from u to v for all (u,v) in V.

The most common algorithm for the all-pairs problem is the floyd-warshall algorithm. This algorithm returns a matrix of values M, where each cell M_{ij} is the distance of the shortest path from vertex i to vertex j. Path reconstruction is possible to find the actual path taken to achieve that shortest path, but it is not part of the fundamental algorithm.

Algorithms

Bellman-Ford Algorithm

The Bellman-Ford algorithm solves the single-source problem in the general case, where edges can have negative weights and the graph is directed. If the graph is undirected, it will have to modified by including two edges in each direction to make it directed. Bellman Ford's algorithm is used to find the shortest paths from the source vertex to all other vertices in a weighted graph.

Bellman-Ford has the property that it can detect negative weight cycles reachable from the source, which would mean that no shortest path exists. If a negative weight cycle existed, a path could run infinitely on that cycle, decreasing the path cost to $-\alpha$.

If there is no negative weight cycle, then Bellman-Ford returns the weight of the shortest path along with the path itself.

It depends on the concept that shortest path has at most n−1 edges as shortest path couldn't have a cycle. In this, there is no need to pass a vertex again since the shortest path to all other vertices could be found without the need for a second visit for any vertices.

Algorithm Steps

➢ The outer loop traverses from 0: n−1.

➢ Loop over all edges, check if the next node distance > current node distance + edge weight, in this case update the next node distance to "current node distance + edge weight".

This algorithm depends on the relaxation principle where the shortest distance for all vertices is gradually replaced by more accurate values until eventually reaching the optimum solution. In the beginning all vertices have a distance of "Infinity", but only the distance of the source vertex = 0, then update all the connected vertices with the new distances, then apply the same concept for the new vertices with new distances and so on.

Implementation:
Assume the source node has a number (0):

```
vector <int> v [2000 + 10];
int dis [1000 + 10];
for(int i = 0; i < m + 2; i++){
v[i].clear();
dis[i] = 2e9;
}
for(int i = 0; i < m; i++){
scanf("%d%d%d", &from, &next, &weight);
v[i].push_back(from);
v[i].push_back(next);
v[i].push_back(weight);
}
dis[0] = 0;
for(int i = 0; i < n - 1; i++){
int j = 0;
while(v[j].size() != 0){
if(dis[ v[j][0] ] + v[j][2] < dis[ v[j][1] ] ){
dis[ v[j][1] ] = dis[ v[j][0] ] + v[j][2];
}
j++;
}
}
```

A very important application of Bellman Ford is to check if there is a negative cycle in the graph.

Dijkstra's Algorithm

Dijkstra's algorithm makes use of breadth-first search to solve the single-source problem. It does place one constraint on the graph which shows that there can be no negative weight edges. For such constraint, Dijkstra greatly improves runtime of Bellman-Ford. Dijkstra's algorithm is sometimes used to solve pairs of shortest path problem by running it on all vertices in V.

Algorithm Steps:

 ➢ Set all vertices distances = infinity except for the source vertex, set the source distance = 0.

 ➢ Push the source vertex in a min-priority queue in the form (distance, vertex), as the comparison in the min-priority queue will be according to vertices distances.

➤ Pop the vertex with the minimum distance from the priority queue.

➤ Update the distances of the connected vertices to the popped vertex in case of "current vertex distance + edge weight < next vertex distance", then push the vertex

➤ with the new distance to the priority queue.

➤ If the popped vertex is visited before, just continue without using it.

➤ Apply the same algorithm again until the priority queue is empty.

Implementation:

Assume the source vertex = 1

```
#define SIZE 100000 + 1
vector < pair < int, int > > v [SIZE]; // each vertex has all the connected vertices with the edges weights
int dist [SIZE];
bool vis [SIZE];
void dijkstra(){
// set the vertices distances as infinity
memset(vis, false, sizeof vis); // set all vertex as unvisited
dist[1] = 0;
multiset < pair < int, int > > s; // multiset do the job as a min-priority queue
s.insert({0, 1}); // insert the source node with distance = 0
while(!s.empty()){
pair <int, int> p = *s.begin(); // pop the vertex with the minimum distance
s.erase(s.begin());
int x = p.s; int wei = p.f;
if( vis[x] ) continue; // check if the popped vertex is visited before
vis[x] = true;
for(int i = 0; i < v[x].size(); i++){
int e = v[x][i].f; int w = v[x][i].s;
if(dist[x] + w < dist[e] ){ // check if the next vertex distance could be minimized
dist[e] = dist[x] + w;
s.insert({dist[e], e} ); // insert the next vertex with the updated distance
}
}
}
}
```

Example 2.3

From the graph, find the shortest path from V3 to all other vertices.

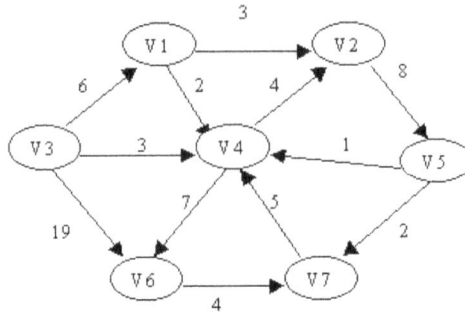

Consider initial data and use the table row which is shadowed:

	D (Distance table)	T	A (Adjacency lists with weights in brackets)	L
V1	-1	-	V1	V2{3}, V4{2}
V2	-1	-	V2	V5 {8}
V3	0	-	V3	V1{6}, V4{3}, V6{19}
V4	-1	-	V4	V2{4}, V6{7}
V5	-1	-	V5	V4{1}, V7{2}
V6	-1	-	V6	V7{4}
V7	-1	-	V7	V4{5}

Further on, DT(i,j), denote the cell in ith row and j-th column of Distance table.

1. Store V3:0 in priority queue (V:n is the distance from V3 to vertex V is **n**)

    PQ: V3:0

2. While queue is not empty, do the following:
    - DeleteMin a vertex from the queue: V3.

      The adjacency list for V3 is:

      AL(3) = V1{6}, V4{3}, V6{19}

    For each ancestor in AL(3), do the following:
    - Read a vertex from the list: V1{6}

      Is distance to V1 calculated?

      No.

      Then do:

      a DT(1,1)← distance to V3 (in DT(3,1)) + edge V3-V1. Now DT(1,1) = 6

a DT(1,2) ← V3 // *record the path*
a Store V1:6 in PQ (the priority queue)
New PQ: V1:6

➢ Read next vertex from AL(3) - V4{3}

The distance to V4 is not calculated, so do the following:
a DT(4,1) ← distance to V3 (in DT(3,1)) + edge V3-V4. Now DT(4,1) is 3.
a DT(4,2) ← V3 // *record the path*
a Store V4:3 in PQ

New PQ: V4:3, V1:6

➢ Read next vertex from AL(3) - V6{19}

The distance to V6 is not calculated, so we do the following:
a DT(6,1) ← distance to V3 (in DT(3,1)) + edge V3-V6. Now DT(6,1) is 19.
a DT(6,2) ← V3 // *record the path*
a Store V6:19 in PQ

New PQ: V4:3, V1:6, V6:19

Distance table and queue after processing the adjacency list of V3:

	DT (Distance table)	
V1	6	V3
V2	-1	-
V3	0	-
V4	3	V3
V5	-1	-
V6	19	V3
V7	-1	-

PQ: V4:3, V1:6, V6:19

Now it is observed that the elements in queue are listed in sorted order. The DeleteMin operation will always read the leftmost element.

➢ **DeleteMin** a vertex from the queue: V4:3

AL(4) = V2{4}, V6{7}
For each ancestor in AL(4) - the adjacency list of V4, we do the following:

➢ Read a vertex from the list: V2{4}

The distance to V2 is not calculated, so do the following:
a DT(2,1) ← distance to V4 (in DT(4,1)) + edge V4-V2. Now DT(2,1) = 7
a DT(2,2) ← V4 // *record the path*
a Store V2:7 in PQ (the priority queue)
New PQ: V1:6, V2:7, V6:19

➢ Read next vertex from AL(4) - V6{7}

The distance to V6 is calculated as 19, which is greater than distance to V4 plus the edge V4-V6, and it has to be **updated**:

a DT(6,1) ← distance to V4 (in DT(4,1)) + edge V4-V6. Now DT(6,1) = 10
a DT(6,2) ← V4 // *record the path*
a Update the priority of V6 in the priority queue

New PQ: V1:6, V2:7, V6:10

Distance table and the queue after processing the adjacency list of V4:

	DT (Distance table)	
V1	6	V3
V2	7	V4
V3	0	-
V4	3	V3
V5	-1	-
V6	10	V4
V7	-1	-

PQ: V1:6, V2:7, V6:10

➢ Delete Min a vertex from the queue: V1:6

AL(1) = V2{3}, V4{2}
For each ancestor in AL(1), adjacency list of V1, do the following:

➢ Read a vertex from the list: V2{3}

The distance to V2 is calculated so far as 7, which is less than the distance to V1 plus the edge V1-V2 (6+3), so do nothing here.

➢ Read next vertex from AL(1) - V4{2}

The distance to V4 is calculated so far as 3, which is less than the distance to V1 plus the edge V1-V4 (6 + 2), so do nothing here.

After processing the adjacency list of V1, the Distance table has not changed, so the priority queue is **V2:7, V6:10**

➢ Delete Min a vertex from the queue: V2:7

AL(2) = V5{8}
For each ancestor in AL(2) we do the following:

➢ Read a vertex from the list: V5{8}

The distance to V5 is not calculated, so do the following:
a DT(5,1) ← distance to V2 (in DT(2,1)) + edge V2-V5. Now DT(5,1) = 15
a DT(5,2) ← V2 // *record the path*
a Store V5:15 in PQ
New PQ: V6:10, V5:15

Distance table and queue after processing the adjacency list of V2 is shown:

	DT (Distance table)	
V1	6	V3
V2	7	V4
V3	0	-
V4	3	V3
V5	15	V2
V6	10	V4
V7	-1	-

PQ: V6:10, V5:15

> Delete Min a vertex from the queue: V6:10
  AL(6) = V7{4}
  For each ancestor in AL(6), the adjacency list of V6, so do the following:
> Read a vertex from the list: V7{4}

  The distance to V7 is not calculated, so do the following:
  a DT(7,1) ← distance to V6 (in DT(6,1)) + edge V6-V7. Now DT(7,1) = 14
  a DT(7,2) ← V6 // *record the path*
  a Store V7:14 in PQ
  New PQ: V7:14, V5:15

Distance table and queue after processing the adjacency list of V6 is shown:

	DT (Distance table)	
V1	6	V3
V2	7	V4
V3	0	-
V4	3	V3
V5	15	V2
V6	10	V4
V7	14	V6

PQ: V7:14, V5:15

> DeleteMin a vertex from the queue: V7:14
  AL(7) = V4{5}
  For each ancestor in AL(7), so do the following:
> Read a vertex from the list: V4{5}

  New distance = distance to V7 plus the edge V7-V4 = 14 + 5 = 19.
  Old distance to V4 computed so far is 3, less than 19.
  We do nothing.

The table has not changed, so the queue will be **V5:15**

➤ **DeleteMin** a vertex from the queue: V5:15

AL(5) = V4{1}, V7{2}

For each ancestor in AL(5), so do the following:

➤ Read a vertex from the list: V4{1}

New distance to V4 is distance to V5 with the edge V5-V4 = 15 + 1 = 16. Old distance to V4 is calculated so far as 3 which is less than 16, so nothing is done here.

➤ Read next vertex from the list: V7{2}

New distance to V7 is distance to V5 plus the edge V5-V7 = 15 + 2 = 17. Old distance to V7 is calculated so far as 14, which is less than 17, so nothing is done here.

The priority queue is empty. The shortest path from V3 to any other vertex is shown:

	D (Distance table)	T	Full path:
V1	6	V3	V3 V1
V2	7	V4	V3 V4 V2
V3	0	-	
V4	3	V3	V3 V4
V5	15	V2	V3 V4 V2 V5
V6	10	V4	V3 V4 V6
V7	14	V6	V3 V4 V6 V7

Topological Sort

For graphs that are directed acyclic graphs (DAGs), a very useful tool emerges for finding shortest paths. By performing a topological sort on the vertices in the graph, the shortest path problem becomes solvable in linear time.

A topological sort is an ordering all of the vertices such that for each edge in, comes before in the ordering. In a DAG, shortest paths are always well defined because even if there are negative weight edges, there can be no negative weight cycles.

Floyd-Warshall Algorithm

The Floyd-Warshall algorithm solves the all-pairs shortest path problem. It uses a dynamic programming approach to do so. Negative edge weight may be present for Floyd-Warshall.

Floyd-Warshall takes advantage of the following observation: the shortest path from A to C is either the shortest path from A to B plus the shortest path from B to C or it's the shortest path from A to C that's already been found. This may seem trivial, but it's what allows Floyd-Warshall to build shortest paths from smaller shortest paths, in the classic dynamic programming way.

If the vertices of a graph G are indexed by {1, 2, ..., n}, then consider a subset of vertices {1, 2, ..., k}. Assume p is a minimum weight path from vertex i to vertex j whose intermediate vertices are drawn from the subset {1, 2, ..., k}. On considering vertex k on path, then either:

k is **not** an intermediate vertex of p, so all intermediate vertices are in {1, 2, ..., k-1}, k is an intermediate vertex of p so,

> ➢ divide p at k giving two sub paths p_1 and p_2 giving $v_i \rightsquigarrow k \rightsquigarrow v_j$
> ➢ as Lemma 24.1, subpaths of shortest paths are also shortest paths where subpaths p_1 and p_2 are shortest paths with intermediate vertices in {1, 2, ..., k-1}

So on defining a quantity $d^{(k)}_{ij}$ as minimum weight of path from vertex i to vertex j with intermediate vertices drawn from set {1, 2, ..., k}, then above properties gives the following recursive solution:

$$d^{(k)}_{ij} = \begin{cases} w_{ij} & \text{if } k = 0 \\ \min\left(d^{(k-1)}_{ik}, d^{(k-1)}_{ij}\right) + d^{(k-1)}_{kj} + d^{(k-1)}_{kj}) & \text{if } k \geq 1 \end{cases}$$

So the optimal values when $k = n$ in a matrix will be:

$$D^{(n)} = \left\{d^{(n)}_{ij}\right\} = \left\{\delta(i, j)\right\}$$

Example 2.4

Consider the directed graph as shown.

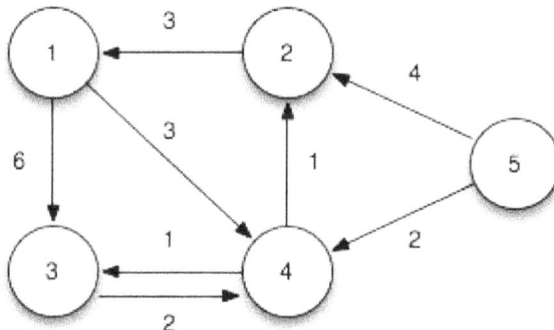

Initialization: *(k = 0)*

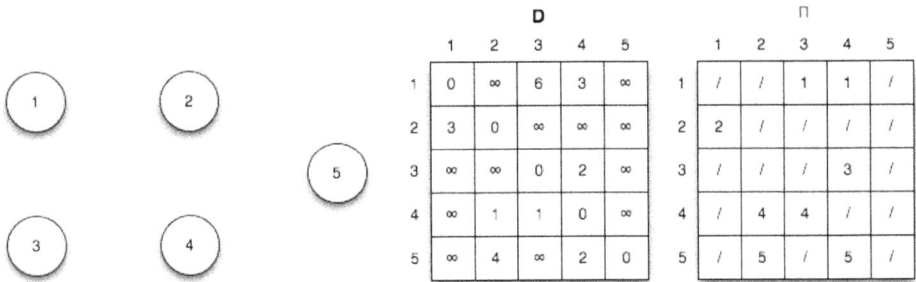

D

	1	2	3	4	5
1	0	∞	6	3	∞
2	3	0	∞	∞	∞
3	∞	∞	0	2	∞
4	∞	1	1	0	∞
5	∞	4	∞	2	0

Π

	1	2	3	4	5
1	/	/	1	1	/
2	2	/	/	/	/
3	/	/	/	3	/
4	/	4	4	/	/
5	/	5	/	5	/

Iteration 1: (k = *1*) Shorter paths from 2 ↝ 3 and 2 ↝ 4 are found through vertex 1

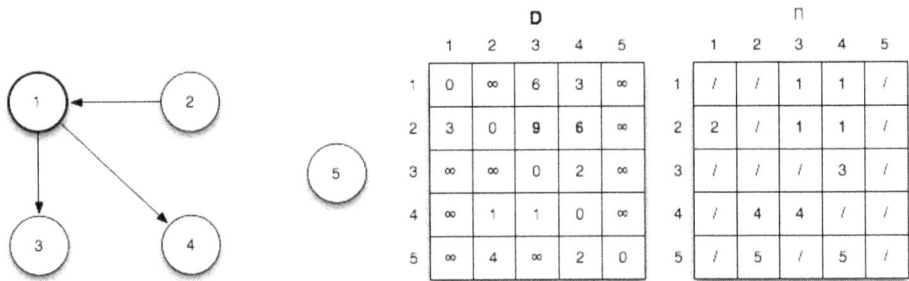

D

	1	2	3	4	5
1	0	∞	6	3	∞
2	3	0	9	6	∞
3	∞	∞	0	2	∞
4	∞	1	1	0	∞
5	∞	4	∞	2	0

Π

	1	2	3	4	5
1	/	/	1	1	/
2	2	/	1	1	/
3	/	/	/	3	/
4	/	4	4	/	/
5	/	5	/	5	/

Iteration 2: (k = *2*) Shorter paths from 4 ↝ 1, 5 ↝ 1, and 5 ↝ 3 are found through vertex 2

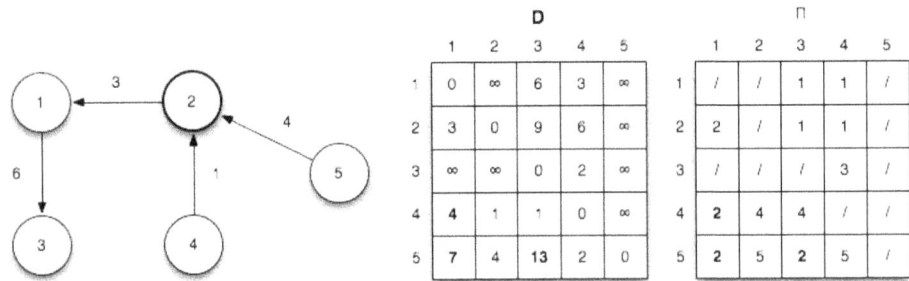

D

	1	2	3	4	5
1	0	∞	6	3	∞
2	3	0	9	6	∞
3	∞	∞	0	2	∞
4	4	1	1	0	∞
5	7	4	13	2	0

Π

	1	2	3	4	5
1	/	/	1	1	/
2	2	/	1	1	/
3	/	/	/	3	/
4	2	4	4	/	/
5	2	5	2	5	/

Iteration 3: (*k* = 3) No shorter paths are found through vertex 3

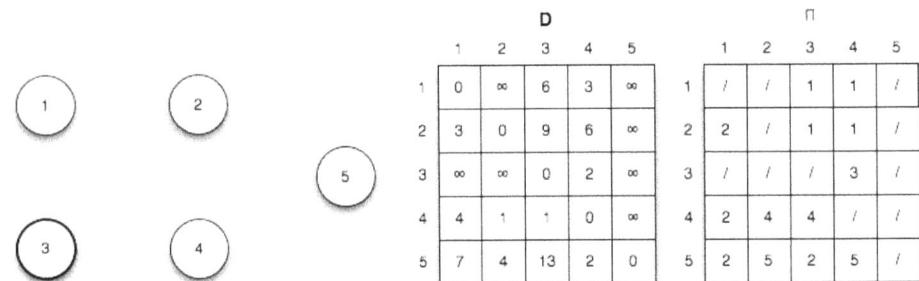

D

	1	2	3	4	5
1	0	∞	6	3	∞
2	3	0	9	6	∞
3	∞	∞	0	2	∞
4	4	1	1	0	∞
5	7	4	13	2	0

Π

	1	2	3	4	5
1	/	/	1	1	/
2	2	/	1	1	/
3	/	/	/	3	/
4	2	4	4	/	/
5	2	5	2	5	/

Iteration 4: ($k = 4$) Shorter paths from $1 \leadsto 2$, $1 \leadsto 3$, $2 \leadsto 3$, $3 \leadsto 1$, $3 \leadsto 2$, $5 \leadsto 1$, $5 \leadsto 2$, $5 \leadsto 3$, and $5 \leadsto 4$ are found through vertex 4

D

	1	2	3	4	5
1	0	4	4	3	∞
2	3	0	7	6	∞
3	6	3	0	2	∞
4	4	1	1	0	∞
5	6	3	3	2	0

Π

	1	2	3	4	5
1	/	4	4	1	/
2	2	/	4	1	/
3	2	4	/	3	/
4	2	4	4	/	/
5	2	4	4	5	/

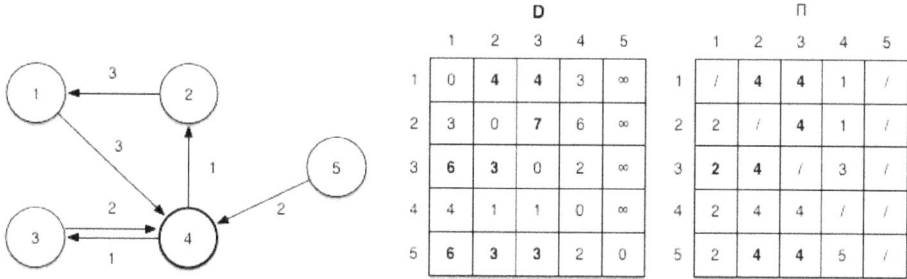

Iteration 5: ($k = 5$) No shorter paths are found through vertex 5

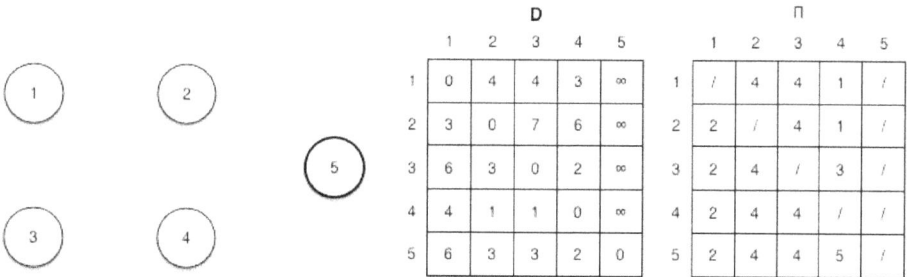

D

	1	2	3	4	5
1	0	4	4	3	∞
2	3	0	7	6	∞
3	6	3	0	2	∞
4	4	1	1	0	∞
5	6	3	3	2	0

Π

	1	2	3	4	5
1	/	4	4	1	/
2	2	/	4	1	/
3	2	4	/	3	/
4	2	4	4	/	/
5	2	4	4	5	/

The final shortest paths for all pairs is given by

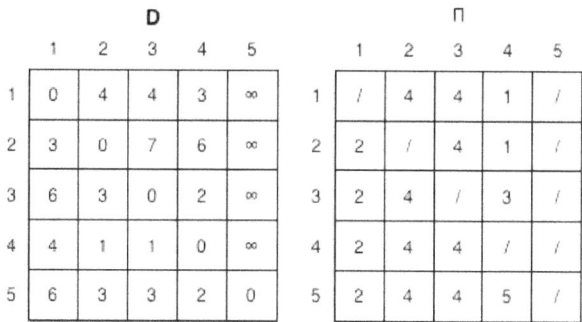

D

	1	2	3	4	5
1	0	4	4	3	∞
2	3	0	7	6	∞
3	6	3	0	2	∞
4	4	1	1	0	∞
5	6	3	3	2	0

Π

	1	2	3	4	5
1	/	4	4	1	/
2	2	/	4	1	/
3	2	4	/	3	/
4	2	4	4	/	/
5	2	4	4	5	/

Johnson's Algorithm

While Floyd-Warshall works well for dense graphs, Johnson's algorithm works best for sparse graphs. In sparse graphs, Johnson's algorithm has a lower asymptotic running time compared to Floyd-Warshall. Johnson's algorithm takes advantage of the concept of reweighting, and it uses Dijkstra's algorithm on many vertices to find the shortest path once it has finished reweighting the edges.

2.3.4 Relation between Maximum Flow and Shortest Path

Suppose we have a directed network G = (V, E) defined by a set V of nodes (or vertexes) and a set E of arcs (or edges). Each arc (i,j) in E has an associated nonnegative capacity u_{ij}. Also we distinguish two special nodes in G: a *source* node s and a *sink* node t. For each i in V we denote by E(i) all the arcs emanating from node i. Let U = max u_{ij} by (i,j) in E. Let us also denote the number of vertexes by n and the number of edges by m.

We wish to find the maximum flow from the source node s to the sink node t that satisfies the arc capacities and mass balance constraints at all nodes. Representing the flow on arc (i,j) in E by x_{ij} we can obtain the optimization model for the maximum flow problem:

Maximise $f(x) = \sum x_{ij}$

Subject to $\sum x_{ij} - \sum x_{ij} = 0$ i ε V \ {s,t}

$0 \leq x_{ij} \leq u_{ij}$

Vector (x_{ij}) which satisfies all constraints is *a* feasible solution or, *a flow*. Given a flow x we are able to construct the residual network with respect to this flow according to the following intuitive idea. Suppose that an edge (i,j) in E carries x_{ij} units of flow, so residual capacity of the edge (i,j) will be $r_{ij} = u_{ij} - x_{ij}$. Hence, additional r_{ij} units of flow from vertex i to vertex j can be send which can cancel the existing flow x_{ij} on arc if x_{ij} units of flow from j to i over the arc (i,j) is send.

So, given a feasible flow x we define the residual network with respect to the flow x as follows. Suppose we have a network G = (V, E). A feasible solution x engenders a new (residual) network, which we define by $G_x = (V, E_x)$, where E_x is a set of residual edges corresponding to the feasible solution x.

2.4 Multi Commodity Flows

In an example shown below, it is shown that how on using decomposition algorithm, minimum-cost multi commodity flow in a directed network is obtained. Such type of problem was motivation for development of original Dantzig-Wolfe decomposition method.

If G = (N,A) is a directed graph, and k be a set of commodities, then for each link (i,j)εA and each commodity k associate a cost per unit of flow, designated by c_{ij}^k then demand/supply at each node i ε N for commodity k is designated as b_i^k, where $b_i^k \geq 0$ shows supply node and $b_i^k < 0$ shows demand node. A decision variables x_{ij}^k shows the amount of commodity k sent from node *i* and node *j* where amount of total flow across all commodities that can be sent across each link is bounded above as u_{ij}.

The problem can be modeled as linear programming problem as:

$$\text{minimize} \quad \sum_{k \in K} \sum_{(i,j) \in A} c_{ij}^{k} x_{ij}^{k}$$

$$\text{subject to} \quad \sum_{k \in K} x_{ij}^{k} \leq u_{ij} \quad (i,j) \in A \qquad \text{(capacity)}$$

$$\sum_{(i,j) \in A} c_{ij}^{k} - \sum_{(j,i) \in A} c_{ij}^{k} = b_{i}^{k} \quad i \in N, \, k \in K \qquad \text{(balance)}$$

$$x_{ij}^{k} \geq 0 \quad (i,j) \in A, \, k \in K$$

Here, constraints limit the total flow across all commodities on each arc which make sure that the flow of commodities leaving each supply node and entering each demand node are balanced.

Consider the directed graph (Figure 2.14).

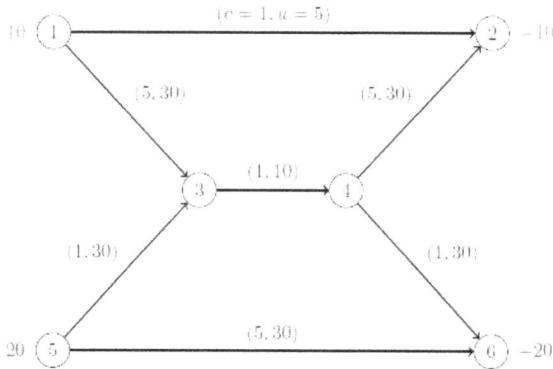

Figure 2.14: Directed Graph.

The idea of such example is to minimize the total cost of sending two commodities across the network while satisfying all supplies and demands and respecting arc capacities. If there were no arc capacities linking the two commodities, you could solve a separate minimum-cost network flow problem for each commodity one at a time.

The following data set *arc_comm_data* provides the cost c_{ij}^{k} of sending a unit of commodity k along arc *(i,j):*

```
data arc_comm_data;
input k i j cost;
datalines;
1 1 2 1
```

```
1 1 3 5
1 5 3 1
1 5 6 5
1 3 4 1
1 4 2 5
1 4 6 1
2 1 2 1
2 1 3 5
2 5 3 1
2 5 6 5
2 3 4 1
2 4 2 5
2 4 6 1
;
```

Next, the data set *arc_data* provides the capacity *uij* for each arc:

```
data arc_data;
input i j capacity;
datalines;
1 2 5
1 3 30
5 3 30
5 6 30
3 4 10
4 2 30
4 6 30
;
```

Lastly, the data set *supply_data* provides the nonzero supply (or demand) b_k^i for each node and each commodity:

```
data supply_data;
 input k i supply;
 datalines;
1 1 10
1 2 -10
2 5 20
2 6 -20
;
```

The following PROC OPTMODEL statements find the minimum-cost multicommodity flow:

```
proc optmodel;
set <num,num,num> ARC_COMM;
num cost {ARC_COMM};
read data arc_comm_data into ARC_COMM=[i j k] cost;

set ARCS = setof {<i,j,k> in ARC_COMM} <i,j>;
set COMMODITIES = setof {<i,j,k> in ARC_COMM} k;
```

set NODES = union {<i,j> in ARCS} {i,j};

num capacity {ARCS};
read data arc_data into [i j] capacity;

num supply {NODES, COMMODITIES} init 0;

read data supply_data into [i k] supply;

var Flow {<i,j,k> in ARC_COMM} >= 0;

min TotalCost =

sum {<i,j,k> in ARC_COMM} cost[i,j,k] * Flow[i,j,k];

con BalanceCon {i in NODES, k in COMMODITIES}:

sum {<(i),j,(k)> in ARC_COMM} Flow[i,j,k]

- sum {<j,(i),(k)> in ARC_COMM} Flow[j,i,k] = supply[i,k];

con CapacityCon {<i,j> in ARCS}:

sum {<(i),(j),k> in ARC_COMM} Flow[i,j,k] <= capacity[i,j];

Because each (balance) constraint involves variables for only one commodity, a decomposition by commodity is a natural choice. In both the OPTLP and OPTMILP procedures, the block identifiers must be consecutive integers starting from 0. In PROC OPTMODEL, the block identifiers only need to be numeric. The following FOR loop populates the .block constraint suffix with block identifier k–1 for commodity *k*:

for{i in NODES, k in COMMODITIES}

BalanceCon[i,k].block = k - 1;

The .block constraint suffix for the linking (capacity) constraints is left missing, so these constraints become part of the master problem.

The following SOLVE statement uses the DECOMP= option to invoke the decomposition algorithm:

solve with LP / presolver=none decomp=() subprob=(algorithm=nspure);

print Flow;

quit;

Here, the PRESOLVER=NONE option is used, because otherwise the presolver solves this small instance without invoking any solver. Because each subproblem is a pure network flow problem, you can use the ALGORITHM=NSPURE option in the SUBPROB= option to request that a network simplex algorithm for pure networks be used instead of the default algorithm, which for linear programming subproblems is primal simplex.

It turns out for this example that if you specify METHOD=NETWORK (instead of the default METHOD=USER) in the DECOMP= option, the network extractor finds the same blocks, one per commodity. To invoke the METHOD=NETWORK option, simply change the SOLVE statement as follows:

solve with LP / presolver=none decomp=(method=network);

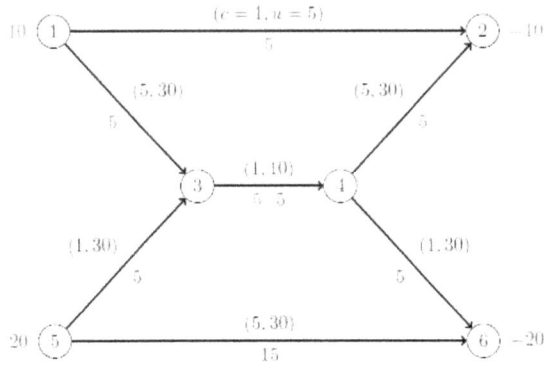

Figure 2.15: Optimal Flow on Network with Two Commodities.

3 | Synthesis of Networks

3.1 Synthesis vs. Analysis

Network analysis is a structured technique used to mathematically analyze a circuit. Quite often the technician or engineer will encounter circuits containing multiple sources of power or component configurations which defy simplification by series/parallel analysis techniques. In those cases, he or she will be forced to use other means. Network analysis starts with a network and by applying the various electric circuit theorems predicts the response of the network. Network synthesis on the other hand, starts with a desired response and its methods produce a network that outputs, or approximates to, that response.

Network synthesis was originally intended to produce filters of the kind formerly described as "wave filters" but now usually just called filters. That is, filters whose purpose is to pass waves of certain wavelengths while rejecting waves of other wavelengths. Network synthesis starts out with a specification for the transfer function of the filter, H(s), as a function of complex frequency, s. This is used to generate an expression for the input impedance of the filter (the driving point impedance) which then, by a process of continued fraction or partial fraction expansions results in the required values of the filter components. In a digital implementation of a filter, H(s) can be implemented directly.

The advantages of the method are best understood by comparing it to the filter design methodology that was used before it, the image method. The image method considers the characteristics of an individual filter section in an infinite chain of identical sections. The filters produced by this method suffer from inaccuracies due to the theoretical termination impedance, the image impedance, not generally being equal to the actual termination impedance. With network synthesis filters, the terminations are included in the design from the start. The image method also requires a certain amount of experience on the part of the designer. The designer must first decide how many sections and of what type should be used, and then after calculation, will obtain the transfer function of the filter. This may not be what is required and there can be a number of iterations. The network synthesis method, on the other hand, starts out with the required function and generates as output the sections needed to build the corresponding filter.

In general, the sections of a network synthesis filter are of identical topology but different component values are used in each section. By contrast, the structure of an image filter has identical values at each section, as a consequence of the infinite chain approach, but may vary the topology from section to section to achieve various desirable characteristics. Both methods make use of low-pass prototype filters followed by frequency transformations and impedance scaling to arrive at the final desired filter.

3.2 Network Topology

Another way to classify computer networks is based on the underlying topology used for constructing the networks. Topology is defined as the geometrical arrangement of nodes. Node are the various computer resources and communication devices (Figure 3.1).

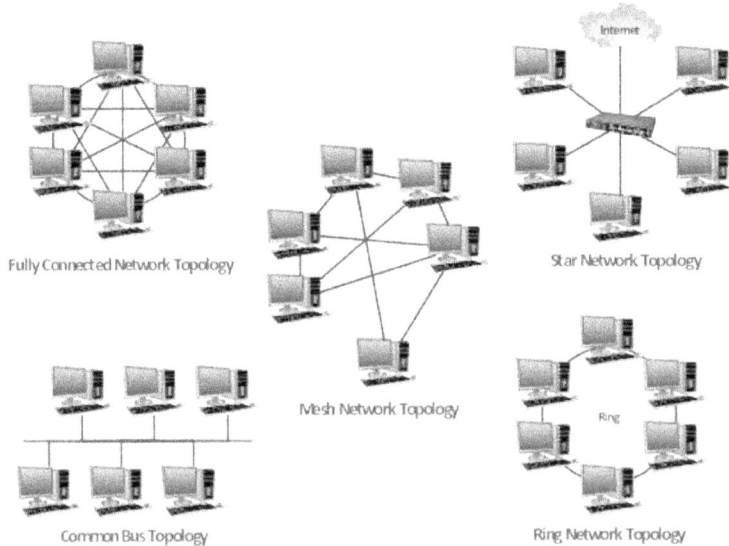

Figure 3.1: Network Topology.

Different Types of Topologies

Following are the different classes of network based on the topological structure.

➤ Bus Network

➤ Star Network

➤ Ring Network

➤ Mesh Network

➤ Tree Network

Bus Network

In a bus network, all nodes are connected to one line known as bus. it is conjointly referred as a time-shared bus. The bus permits just one pair of nodes to establish communication at a time. This property restricts the total number of nodes connected to form a reliable bus network. However, several protocols were developed for a bus to form communication more efficient and reliable. CSMA/CD and Token bus protocols ar sensible examples. The structure of a bus network is shown in Figure.

Advantage of a bus network is its ability to connect any number of nodes without extensive hardware. Nodes can also be removed from the bus simply. it's straightforward to maintain the bus network (Figure 3.2).

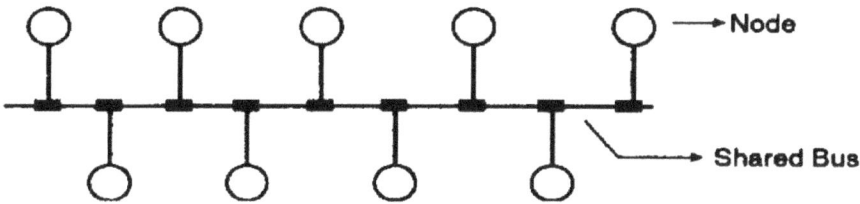

Figure 3.2: Bus Network.

Star Network

In a star network, each node is connected by means of a dedicated point-to-point(P2P) channel to a central node called server that will act as a switch. The central server will provide the connectivity for all pair of nodes willing to communicate with each other. But, if the central server fails, the whole network will also fail. The transmission media may be a twisted pair, coaxial cable or optical fibre. Structure of a star network is shown in Figure 3.3.

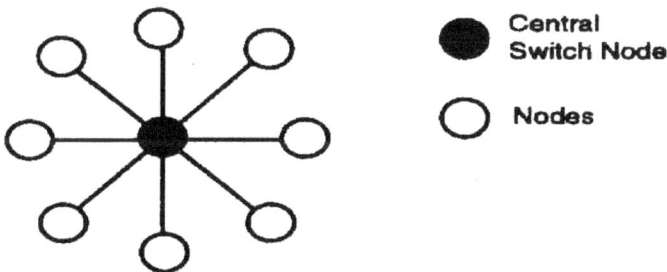

Figure 3.3: Star Network.

Some of the advantages of star network are:

> ➢ Easy implementation
>
> ➢ Centralized control
>
> ➢ Simple access protocols

The main disadvantage of star network is that they suffer from the problem of central node failure. They also require long cable length; each new device requires an exclusive cable. Campus PBXs are often implemented using star network topology.

Ring Network

Nodes in a ring network are connected in the type of a closed loop. One communication channel is commonly implemented to provide the connectivity. Data from the sending node circulates round the ring till it reaches the destination. a ring will be unidirectional or bi-directional. In a unidirectional ring, data moves in one direction solely. In a bi-directional ring, data can move in both directions, but moves in one direction at a time. Single node failure may paralyse the transmission of information to a set of nodes in a unidirectional ring. but messages will be sent to nodes in either side of the affected node (Figure 3.4).

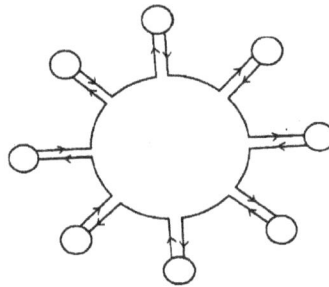

Figure 3.4: Ring Network.

Ring network with a method called token passing (Token Ring) was proposed by IBM and approved by IEEE as one of the standards for LAN. Advantages of a ring network are its short cable length, suitability for optical fibre implementation and its flexibility to include new nodes which is also called as Network expansion. Disadvantages of ring networks include the failure of entire network in the presence of a single node failure, difficulty in diagnosing faults and its non-adaptability to structural changes.

Mesh Network

In a mesh network, each pair of nodes is connected by means of an exclusive point-to-point link. Each node requires a separate interface to connect with the other device. Mesh networks are seldom constructed in practice. They are useful in situations, where one node or station needs to frequently send messages to all other nodes. Otherwise, a considerable amount of network bandwidth got wasted. The advantages of mesh network are excessive amount of bandwidth and inherent fault-tolerance. The structure of a mesh network is shown in Figure 3.5.

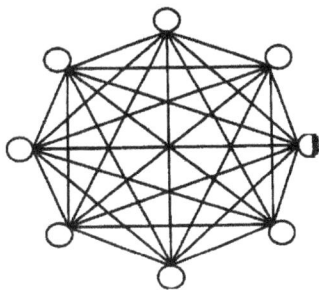

Figure 3.5: Mesh Network.

Tree Network

A tree network is another form of bus network. Several nodes are connected into a hierarchical form as shown in diagram. The root node may be a powerful server or a mainframe computer often called a head-end. Tree networks are suitable for organizations, where head offices need to communicate with regional offices and regional offices needs to communicate with remote offices. Advantages of a tree network are its ease of expansion, identification and isolation of faulty nodes whereas its disadvantage is that, it also suffers from the problem of the network being highly dependent on the root node (Figure 3.6).

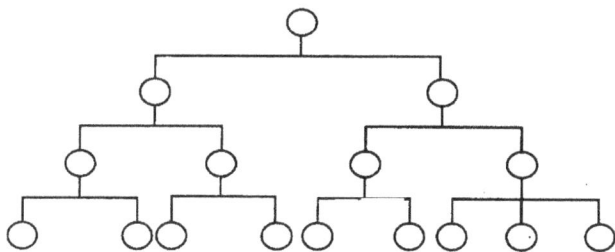

Figure 3.6: Tree Network.

3.3 Elements of Circuit Synthesis

3.3.1 Nullor and Nullor-Mirror Equivalents

Nullor is useful for analysis, synthesis and designing of procedures since it facilitates modelling the behaviour of active device disregarding to its particular realization of active blocks. It is known that nullator norator combinations is difficult to realize CCII+, ICCII+ and ICCII− unless additional resistors are used. Pathological voltage mirror (VM) and pathological current mirror (CM) provides means of showing four single-output CCII and ICCII members without any need of using resistors. Systematic synthesis of active circuits developed is based on using nullor elements, which aims to adjust the pathological mirror elements by allowing more ideal representations of active circuits.

Genetic Algorithm is optimization technique which works on the principle of survival of fittest which has the capability to produce new design solutions from population of current solutions, and discarding the solutions which have inferior performance or fitness. Genetic Algorithm starts from high level descriptions to synthesize analog circuits. The automatic synthesis of analog circuits from high level specifications is located as challenging problem. It is observed that genetic algorithm with nullator based theory was applicable to produce Voltage Followers circuits. Such method shows how an automatic system can deal with huge search spaces to design practical VFs by performing ancient operations from nullator based theory. Further, generation of Voltage Mirrors circuits based on genetic algorithm was presented that introduced new genetic algorithm to synthesis negative type CCII- blocks by superimposing VFs and current followers.

As observed, second order filter using operational amplifiers has been reviewed in which passive and active compensation methods improves circuit performance for high designs. With this, classical TT circuit using op-amps has frequency limitations due to finite gain bandwidth of opamps. Figure 3.7 shows an active-RC topology which is applied to realize lowpass and bandpass biquadratic filtering. These topology has been widely used as it is simple, versatile, ad require few components.

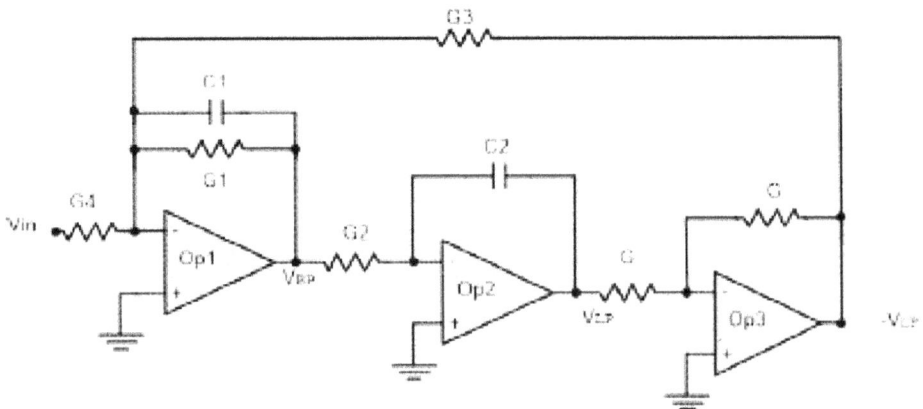

Figure 3.7: Second Order Filter with Three Op Amps.

Method

Genetic Algorithm is carried out in various steps as described in flow chart in the Figure 3.8.

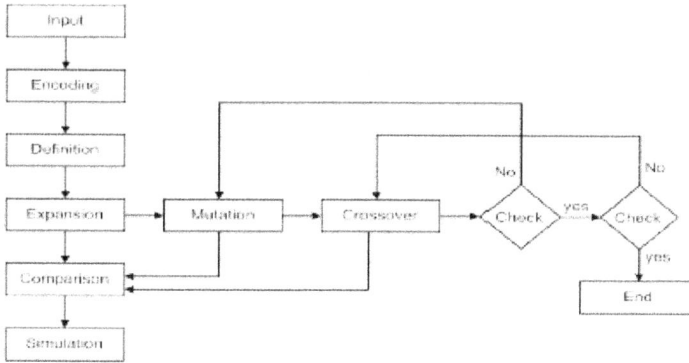

Figure 3.8: Flow Chart for the Proposed Algorithm.

In this, the nodal equation for circuit is written in form [Y][V]=0 so that element will able to count which can be extracted. The number of nonzero elements in Y matrix shows total number of elements of circuit, where a single increment of number of rows shows number of original nodes, number of diagonal elements showing number of grounded elements, and reminder elements appears as floating.

Next, all passive elements are to be encoded in genes and floating passive elements are to be expanded as:

Each element is shown by a gene of form Gen:R·C·S, where R is row number, C is column number, S is sign bit. Sign bit is 1 for positive elements and is 0 for negative elements.

If column and row numbers are same, then element is grounded and is shown for simplicity by Gen:R·0 else an expansion subroutine for element is applicable.

Table 3.1: Nullor and the Pathological Elements

Pathological Element	Symbol	Genetic representation
Nullator		O
Norator		P
Voltage mirror		V
Current mirror		I

Table 3.1 shows the encoding of Nullor and the pathological elements where current conveyors types are considered in CCII+, CCII-, ICCII+, and ICCII- as shown. Such current conveyors can be codified in genes and is shown in Table 3.2.

Table 3.2: CCII Symbols with Representation

CCII symbol	Nullor and MirrorRepresentation	Genetic Representation
		Z.X.I.X.Y.O
		Z.X.P.X.Y.O
		Z.X.I.X.Y.V
		Z.X.P.X.Y.V

Recently a systematic generation method of TA based on nullor elements and pathological mirror elements was employed to provide pathological realizations of different types of TA. The Balanced Output TA (BO-TA) and Single Output TA (SO-TA) are two different types of TAs. In this, a Single-input-Single Output SISO-TA will be considered having four configurations for SISO-TA. To make possible in proposed algorithm, everyone is codified in genes and shows in Nullor form from Table 3.3.

Table 3.3: Nullor Form

TA symbol	Nullor and Mirror Representation	Genetic Representation
TAI-		Gen | = U.S.P.S.M.O
		Gen | = U.S.I.S.M.V
TAO-		Gen | = U.S.P.S.N.O
		Gen | = U.S.I.S.N.V
TAI+		Gen | = U.S.I.S.M.O
		Gen | = U.S.P.S.M.V
TAO~		Gen | = U.S.I.S.N.O
		Gen | = U.S.P.S.N.V

Gene codification of positive resistances can be obtained using the first and the second configurations while negative resistances can be obtained using the third and fourth configurations.

Further, subroutine for floating elements is applied where elements are expanded based on sign bit (S). If S is 1, it is positive element where gene can be written as: Gen/passive element = R.N.P.N.C.O.OR Gen/passive element -R.N.I.N.C.V. where N is the insertion node.

If S is 0; it is negative element, hence gene can be written as: Gen/passive element R.N.I.N.C.O.OR Gen/passive element R.N.P.N.C.V

The generated genes of passive elements are compared with active counterpart whose equivalent circuits are given in above tables. The generated circuit is tested for optimum ωO and Q. The genes are further arranged together to form chromosome. A mutation operation is applied to each gene in chromosome which resulted from subroutine as described. The mutated genes and the genes of the active elements are also compared. The performance of the generated circuits are also tested for optimum ωO and Q

TA-TT EQUIVALENT CIRCUITS USING TT_GA

A:Analysis:

Here the proposed algorithm is applied to have TA-TT equivalent circuits. The above steps shows that the grounded resistance is realized using TA as shown in Figure 3.9 where realization of figure is used to represent the grounded transconductance G1.

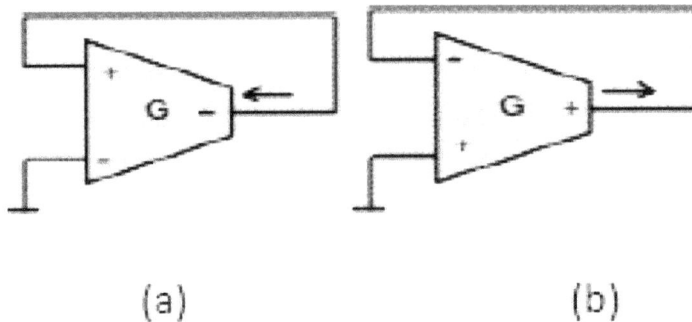

(a)   (b)

Figure 3.9: Representation of Grounded Resistance using TA.

In the next step, the chromosome shows the expanded elements as:

= Gen2 I−G3 . Gen3 I G4 . Gen4 I G2 .

On using first possible to all genes, chromosome will be:

= 1.4.I.4.2.O.1.5.P.5.3.O.2.6.P.6.1.O

If on comparing among genes of SO-TA family representations and resulted chromosome, it is observed that Gen2 can be shown by TAI+ in table above with negative terminal as grounded while positive terminal is connected to node 2 and output terminal is connected to node 1, with Gen3 and Gen4 are shown by TAI- 1in table above.

Hence the final output file will be:

sC2 2 0 sC1 1 0 G1 1 1 TAI+ 1 2 TAI- 1 3 TAI- 2 1

It is observed that the output responses; VBP and VLP are taken from nodes 1 and 2 and input is injected through node 3 which is shown in Figure 3.10.

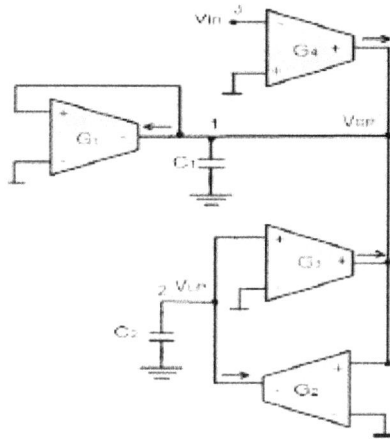

Figure 3.10: SISO-TA Representations of TT_GA Filter.

Since there are 3 floating elements and grounded element that shows TA and every element has two configurations with each configuration shown by two genes hence, 44 = 256 which is different forms of equivalent nullor mirror realizations for filter and 24 = 16 equivalent SO-TA- realizations. These realizations are obtained by applying mutation and crossover among genes. On applying mutation operation to second gene results in new chromosome:

= 1.4.I.4.2.O.1.5.I.5.3.V.2.6.P.6.1.O

After mutation the nullor representation of the circuit change and Gen3 can be represented by the TAI- and TAO- in table III as we use TAI- in the first realization so we will use TAO-. Gen4 is represented by the TAI-and Gen2 can be represented by the TAI+. The final realization of TT_GA with SO-TA is shown in Figure 3.11.

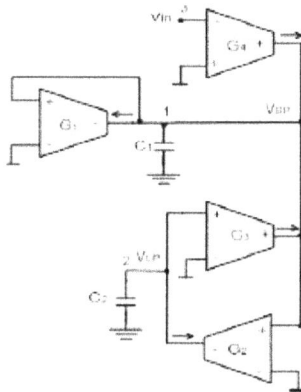

Figure 3.11: SISO-TA Representations of TT_GA Filter.

P-Spice simulations were done to verify the performance of equivalent TA-TT_GA circuits. In this, CMOS realization of TA circuit is used for simulation with 0.25µ model and supply voltages of ±1.5V. TA-TT_GA circuit as shown in figure above is designed to have f_0 =1MHz and Q = 10. The input signal is a sinusoidal input voltage source of 1V magnitude. Figure 3.12 shows the BP and LP magnitude responses of TA-TT_GA configuration of Figure 3.11 which is compared with original op amp TT circuit.

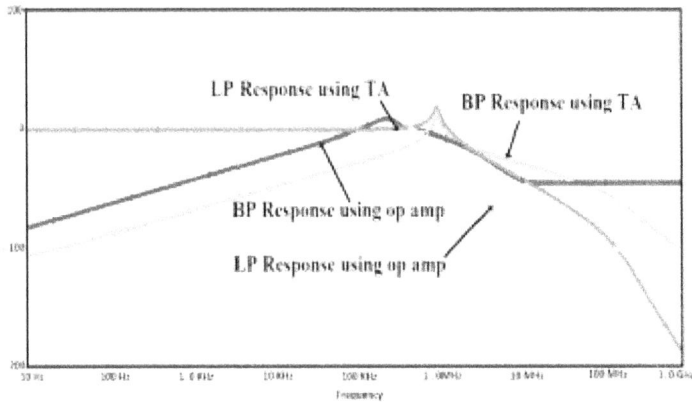

Figure 3.12: Magnitude Response of TA-TT Configuration of
Figure 3.11 Compared with Original Op Amp TT Circuit.

From above TA circuit it is clear that the circuit has superior performance as compared to original TT one. A magnitude response comparison between TA-TT_GA configuration is shown in Figure 3.12 whereas an ideal response is shown in Figure 3.13 where the errors in the centre frequency and quality factor are less than 10% and 5% respectively.

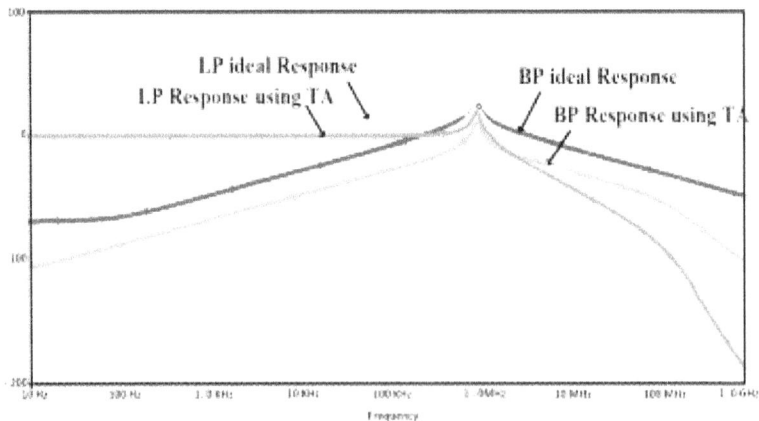

Figure 3.13: Magnitude Response of TA-TT Configuration of
Figure 3.12 Compared with Ideal Response.

3.4 Process Network Synthesis

Process Network Synthesis (PNS) is a method to represent a process structure in a 'directed bipartite graph'. The Process Network Synthesis uses the P-graph method to create a process structure. Scientific aim of this method is to find optimum structures.

Process Network Synthesis uses a bipartite graph method P-graph[1] and employs combinatorial rules to find all feasible network solutions (maximum structure) and links raw materials to desired products related to the given problem. With a branch and bound optimisation routine and by defining the target value an optimum structure can be generated that optimises a chosen target function.

Process Network Synthesis was originally developed to solve chemical process engineering processes. Target value as well as the structure can be changed depending on the field of application. Thus many more fields of application followed.

Process Network Synthesis is used in different applications where it can be used to find optimum process structures like:

> ➢ Process engineering: Chemical process designs and the Synthesis of chemical processes is applied in different case studies.

> ➢ Optimum energy technology networks for regional and urban energy systems: In case of regional and urban energy planning the financially most feasible solution for resource systems is selected as target value. With this setting material- and energy flows, energy demand and cost of technologies are considered and the optimum technology network can be found. Simultaneously the robustness of technologies due to price changes and limitations in resource availability can be identified.

> ➢ Evacuation routes in buildings: The aim is to find optimal routes to evacuate buildings depending on specific side parameters.

> ➢ Transportation routes: In this, area transportation routes with minimum cost and lowest environmental impact can be identified.

Mathematical Programming

Mathematical programming approach to design, integration, operation and synthesis problems comprises of certain major steps.

> ➢ Initially, the development of alternatives from which the optimum solution is selected needs to be considered.

> ➢ The formulation of mathematical program which involves discrete and continuous variables for the selection of configuration and operating levels are considered.

> ➢ The solution of optimization model from which optimal solution is determined.

Significant advances have taken place with this methodology, which offers the possibility of developing automated tools to support the exploration of alternatives and optimization by design engineers. In certain decades, there have been much advances in mathematical programming techniques. In case of solution of mixed-integer nonlinear programming problems and rigorous global optimization of nonlinear programs, there are advances in capability of solving very large problems, particularly for linear and mixed-integer linear programming techniques. There has also been recently a trend towards new logic-based formulations that can facilitate the modeling and solution of these problems. Finally, the availability of modeling systems that can facilitate the formulation of optimization problems has also made great progress, as well as the development of alternative solution strategies.

Design and synthesis problems give rise to discrete/continuous optimization problems, which when represented in algebraic form, correspond to mixed-integer optimization problems that have the following form:

$$\min \; : z = f(x, y)$$
$$s.t \; h(x){=}0$$
$$g(x) = 0$$
$$x \in X \subseteq \mathfrak{R}^{n}$$
$$y \in \{0,1\}^{m}$$

where
f(x, y) = objective function
h(x, y) = 0 which is equations showing performance of a system and heat balances
g(x, y) <= 0 = inequalities showing specifications or constraints for feasible choices

Here variables x are continuous and corresponds to state or design variables while y are the discrete variables that are restricted to take 0-1 values to define selection of item or action. Problem (MIP) corresponds to a mixed-integer nonlinear program (MINLP) when any of the functions involved are nonlinear. If all functions are linear it corresponds to a mixed-integer linear program (MILP). If there are no 0-1 variables (MIP) reduces to a nonlinear program (NLP) or linear program (LP) depending on whether or not the functions are linear.

Normally, formulation and solution of different mathematical programming problems can be effectively performed with modeling systems like GAMS and AMPL as they require model to express explicitly in algebraic form since they have advantage as of being automatically interface with codes for solving types of problems. They perform automatic differentiation and allow the use of indexed equations, with which large scale models can be readily generated. It should also be noted that these modeling systems now run mostly on desktop and PC computers, making their use and application widely available.

The solution of LP problems relies largely on the simplex algorithm, although lately interior-point methods have received increased attention for solving very large problems given their polynomial complexity. MILP methods rely largely on simplex LP-based branch and bound methods that consists of a tree enumeration in which LP subproblems are solved at each node, and eliminated based on bounding properties. These methods are being improved through cutting plane techniques, which produce tighter lower bounds for the optimum. LP and MILP codes are widely available. The best known include CPLEX, GUROBI, OSL and XPRESS, all which have achieved impressive improvements in their capabilities for solving problems. On the other hand, since MILP problems are NP-complete it is always possible to run into time limitations when solving problems with large number of 0-1 variables, especially if the integrality gap is large.

The solution of NLP problems relies either on the successive quadratic programming algorithm (SQP), or on the reduced gradient method. Major codes include MINOS and CONOPT for the reduced gradient method, and OPT or SNOPT for the SQP algorithm. These NLP methods are guaranteed to find the global optimum if the problem is convex. When the NLP is non-convex a global optimum cannot guaranteed. One option is to try to convexify the problem, usually through exponential transformations, although the number of cases that this possible is rather small. Alternatively, one could use rigorous global optimization methods which over the last few years have made significant advances. These methods assume special structures such as bilinear, linear fractional and concave separable. Although this may appear to be quite restrictive, it has be shown that algebraic models are always reducible to these structures, provided they do not involve trigonometric functions.. Computer codes for global optimization still remain in the academic domain, and the best known are BARON and α-BB. It should also be noted that non-rigorous techniques such as simulated annealing and genetic algorithms, which have also become popular, do not make any assumptions on the functions, but then they cannot guarantee rigorous solutions, at least in finite amount of time. Also, these methods do not formulate the problem as a mathematical program since they involve procedural search techniques that in turn require some type of discretization.

Major methods for MINLP problems include first Branch and Bound (BB), which is a direct extension of the linear case, except that NLP sub-problems are solved at each node. Generalized Benders Decomposition (GBD), and Outer-Approximation (OA), are iterative methods that solve a sequence of NLP subproblems with all the 0-1 variables fixed, and MILP master problems that predict lower bounds and new values for the 0-1 variables. The difference between the GBD and OA methods lies in the definition of the MILP master problem; the OA method uses accumulated linearization's of the functions, while GBD uses accumulated Lagrangian functions parametric in the 0-1 variables. The LP/NLP based branch and bound essentially integrates both sub problems within one tree search, while the Extended Cutting Plane Method (ECP) does not solve the NLP subproblems, and relies exclusively on successive linearization's. All these methods assume convexity to guarantee convergence to the global optimum. Non-rigorous methods for handling non-convexities include the equality relaxation algorithm and the augmented penalty version of it. The only commercial code for MINLP is DICOPT (OA-GAMS), although there are a number of academic versions (MINOPT, a-ECP).

It is observed that as per new trend emerged in formulation and solution of discrete/continuous optimization problems through a model known as Generalized Disjunctive Programming explains that it can be applied for boolean and continuous variables which formulate the problem with objective function subject to types of constraints:

a) global inequalities that are independent of discrete decisions

b) disjunctions that are conditional constraints involving an OR operator

c) pure logic constraints involving boolean variables

More specifically the problem is given as follows:

$$\min : z = \sum_{k \in K} c_k + f(\mathrm{x})$$

$$s.t \; \mathrm{h}(\mathrm{x}) = 0$$

$$g(\mathrm{x}) \le 0$$

$$\bigvee_{j \in D_j} \begin{bmatrix} Y_{j,k} \\ h_{j,k}(\mathrm{x}) = 0 \\ g_{i,k}(\mathrm{x}) \le 0 \\ c_{i,j} = \gamma_{i,j} \end{bmatrix} k \in K$$

$$\Omega(Y) = True$$

$$x \in X \subseteq \Re^n ; Y \cup \{True, False\}^m$$

where

x = continuous variables

y = boolean variables

The objective function here has term $f(x)$ for continuous variables and charges ck which depends on discrete choices. The equalities/inequalities $g(x)$ must hold regardless of the discrete conditions and $h_{ik}(x) = 0$ are conditional equations that must be satisfied when the corresponding boolean variable y_{ik} is True for i'th term of k'th disjunction. Also, fixed charge ck is assigned the value g_{ik} for that variable. Hence, the constraints W(y) involve logic propositions in terms of boolean variables.

GDP shows an extension of disjunctive programming, which earlier has been used as a framework for deriving cutting planes for algebraic problem. It is interesting to note that any GDP problem can be reformulated as a MIP problem and vice-versa. It is more natural, however, to start with a GDP model, and reformulate it as a MIP problem. This is accomplished by reformulating the disjunctions using the convex hull transformation or with big-M constraints. The propositional logic statements are reformulated as linear inequalities. For the linear case of problem GDP, and when no logic constraints are involved, Beaumont proposed a branch and bound method that does not rely on 0-1 variables and branches directly on the equations of the disjunctions. This method was shown to outperform the solution of the alternative algebraic MILP models. Raman and Grossmann developed a branch and bound method for solving problem GDP in hybrid form; *i.e.* with disjunctions and mixed-integer constraints.

For nonlinear case of problem and for case of process networks, it is observed that a logic-based Outer-Approximation algorithm which is based on idea of extending Outer-Approximation algorithm by solving NLP sub problems in reduced space is preferred in which constraints that do not apply in disjunctions are disregarded, from which both efficiency and robustness can be improved. In such method, MILP master problems correspond to convex hull of linearization of the nonlinear inequalities takes place. Many NLP sub problems are solved to initialize the master problem in order to cover all the terms in the disjunctions. Such method has been implemented in the computer prototype LOGMIP, a GAMS-based computer code.

There are several solution strategies that can be used in the optimization of mathematical programming models for design and synthesis. The two major strategies are simultaneous optimization, and the sequential optimization. In the simultaneous strategy a single model is optimized at once. The optimization is rigorous in that all the trade-offs are taken simultaneously into account. Also, this model used is commonly of one type, but hybrids are possible. For example one can perform simultaneous optimization of a flow sheet in which the reaction, separation and heat integration are each represented by aggregated models. Alternatively, simultaneous optimization can be applied to the synthesis of subsystems, for example heat exchanger networks, or heat integrated distillation units.

The sequential optimization strategy consists of solving a sequence of sub problems, normally at an increasing level of detail. The major motivation is to solve simpler problems to avoid solving a large single problem. A good example is the procedure implemented in MAGNETS in which an LP is solved first to target the utility cost, next an MILP to determine the identity of the fewest number of matches, and finally an NLP superstructure in which interconnection of exchanger with the predicted matches are determined.

While the simultaneous and sequential strategies have been used for a long time, there are several new variants that have been proposed. For instance, in sequential decomposition one can use aggregated, short-cut and detailed models. In simultaneous optimization one can use a model of a single type, or a mix of several types. Finally, it is possible to use mathematical programming in combination with other approaches, most notably, physical insights.

3.5 Design Methodology for Networked Embedded Systems

Several system level design exploration methodologies exist that help designers to transform a high level specification in to an implementation on a SoC or embedded system.

SoC is System-on-Chip whose products and embedded systems are rapidly growing in complexity. It is observed that an enormous growth of complexity of user requirements, like flexibility, multiple functionality, autonomy and cost, Also, a continuous improvement in the capabilities of underlying hardware where number of transistors per chip increases at a tremendous pace and clock speed improvements of electronic devices are beyond imagination.

Unfortunately the mapping efficiency of applications does not scale. It decreases continuously. As an example: a Pentium-4 chip at 2GHz does not perform linearly better than its 486 ancestor at 66MHz, despite the fact that it contains many more transistors and the power density is significantly higher. For embedded systems, where additional constraints such as the power consumption are very important, this picture looks even worse.

In order to make the problems still more difficult to solve, it is noticed that entering the deep sub-micron technologies (90nm and less) is also making the design community face new challenges: interconnect on these chips is becoming a key bottleneck for the performance of the chips, and managing the leakage of the active devices is becoming essential for power efficient systems.

Despite this example, in the past, system champions were still able to obtain reasonably good results in a reasonable design time. Their success is based on years of experience, extensive know-how and feeling. However, in the future even these champions will need to be assisted by new technologies and established and proven design methodologies, that guide designers through the complex and tedious job of designing complex embedded systems. Another problem that has to be solved is the lack of these system champions: in order to increase the efficiency and reduce the lead-time of these designs, intelligent methodologies supported by the necessary design tools are a must.

Research done at IMEC on design methodology always focused on the development and demonstration of design methods that enable the design of advanced SoC-based products and embedded systems and optimize the design quality and the design productivity. We focus on the trajectory between a given abstract application specification and the mapping on an ASIC or platform architecture, as described in Figure 3.14.

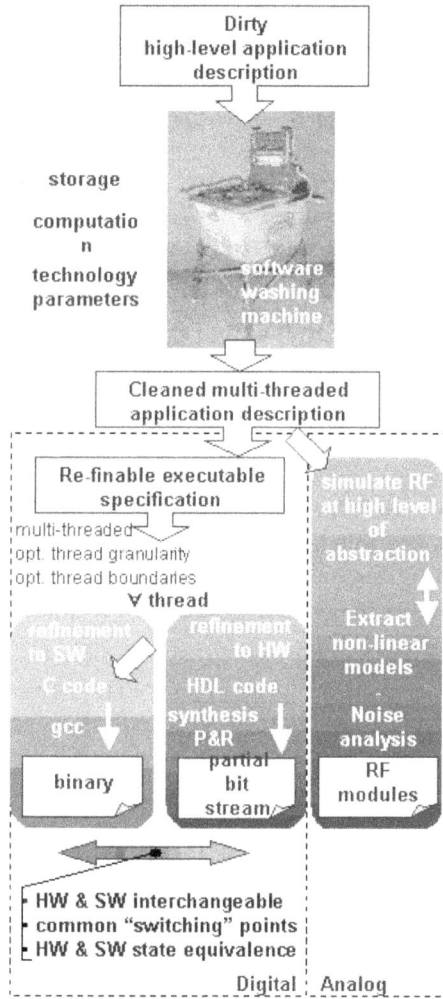

Figure 3.14: Platform Architecture.

Often an application specification is described in an executable language, like C++ or MATLAB, which puts an emphasis on the functionality and the proof of concept of the algorithm. By means of code structure changes, optimized code is generated with emphasis on speed, energy consumption, efficient memory hierarchy and others, and this within constraints like real-time performance and QoS. Cleaning the application description will ultimately result in nicely multi-threaded descriptions that can run on multi-processor platforms, and that have an optimized memory architecture. For each of these threads (fp thread) a designer now has to implement that function. By the insertion of constructs that take concepts of time (registers) and bit-width, inter-processor messages and tokens into account a complete multi-process system can be described and simulated, including processes on hardware accelerators or instruction set processors (ISP¡¦s). Important is that the designer has a continuous view on the correctness of his design by comparing it to a basic specification.

One of the major design challenges for a system designer is the partitioning of the different threads in to software, *i.e.* functions executed on one or more ISP¡¦s, or on hardware, *i.e.* dedicated hardware accelerators. One solution to this partitioning problem is trial and error. A better approach consists of describing the different functional units in a unified language, and of only making the choice between HW and SW at the last moment, when the timed bit-true description of all functional blocks is known.

For a co-design of analogue and digital modules, you need at the high abstraction level a methodology to model and simplify analogue modules in order to efficiently simulate them together with digital modules.

Another problem is the co-habitation of digital and analogue modules on the same die, *i.e.* SoC systems. Digital systems are very robust to parasitic signals, but they create noise in the substrate of the SoC. This noise has a significant influence on the analogue circuits. Being able to model these noise sources, and understanding how this noise is reaching the sensitive analogue circuits allows the designer to make choices and trade-offs that will make his SoC more robust and fault-free.

The design flow described in Figure 3.15 is shown by set of IMEC design technologies which solves some design problems in that area. In next figure, logical chain of such tools is given, some mature, some still in a basic research phase.

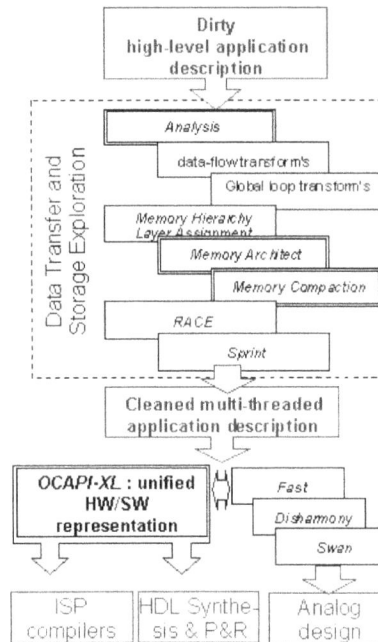

Figure 3.15: The IMEC Tool Flow.

The need to optimize application descriptions at the higher abstraction level for memory access is driven by the fact that for data intensive applications the data transfer and storage related actions are significant power drains.

It is observed that clock speed of processors is increasing significantly over time, but access speed of external memory is lagging behind. The consequence of that is that many applications today are no longer limited in speed by their calculation performance (Gops), but by the access to (common) memory to store and retrieve data. By taking memory hierarchy and data transfer and storage in to account in an early phase, these memory bottlenecks can be removed and better overall design results obtained.

Data flow transformations modifies the algorithmic data-flow to remove any redundant data transfers that are typically present in real-life code. Data flow transformations also serve as enabling transformations for other steps in the global DTSE methodology as they break data-flow bottlenecks.

Global loop transforms mainly increase the locality and regularity of the accesses in the code. In an embedded context this is clearly good for memory size and memory accesses but of course also for performance. So this step should be seen as a way to improve the quality of the code before the more detailed but also quite locally applied conventional compiler/synthesis loop optimizations. This preprocessing also enables later steps in our DTSE script.

Memory Hierarchy Layer Assignment decides how to optimally assign the application data to the various memory hierarchy layers such that data reuse is maximized and costly redundant inter-layer transfers are minimized. It takes the constraints of the ATOMIUM/MA step into account to guarantee a solution that meets the real-time constraints of the application.

The implementation of embedded networked appliances requires a mix of processor cores and HW accelerators on a single chip. When designing such complex and heterogeneous SoCs, the HW/SW partitioning decision needs to be made prior to refining the system description. With OCAPI-xl, we developed a methodology in which the partitioning decision can be made anywhere in the design flow, even just prior to doing code-generation for both HW and SW.

OCAPI-xl is a C++ class library targeting system design for heterogeneous HW/SW architectures. The executable model contains the functionality specified in a target-independent way. It can also contain the architectural properties, but both are specified independently.

During design space exploration, OCAPI-xl gives the designer simulation based performance feedback. This allows exploring multiple HW/SW partitioning alternatives and making a well-founded decision for the final implementation. The path down to implementation is based on incremental refinement and automatic code generation. It results in plain ANSI-C code for the software components and synthesizable register transfer HDL for hardware parts.

3.6 Network Flows in Economy

The circular flow of economic activity is a model showing the basic economic relationships within a market economy. It illustrates the balance between injections and leakages in our economy. Half of the model includes injections, and half of the model includes leakages. The circular flow model shows where money goes and what it's exchanged for. The model includes households, businesses and governments. In circular flow of economy, money is used to purchase goods and services. Goods and services flow through the economy in one direction while money flows in the opposite direction.

The factors of production include land, labor, capital and entrepreneurship. The prices that correspond to these factors of production are rent, wages and profit. People in households buy goods and services from businesses in an attempt to satisfy their unlimited needs and wants. Households also sell their labor, land, and capital in exchange for income that they use to buy goods and services that firms produce. Businesses sell goods and services to households, earning revenue and generating profits. Businesses also pay wages, interest and profits to households in return for the use of their factors of production. Governments levy taxes on households and businesses in order to provide certain benefits to everyone.

4 | Network Simplex Method

4.1 Introduction

Simplex method is an approach to solve linear programming models using slack variables, tableaus, and pivot variables for finding optimal solution of a problem. The Simplex Method normally does a tour of boundary of feasible region stopping at vertices so as to examine the value of objective function. In this, there appears a recognisable situation when it reached the optimal solution where each vertex is not physically examined. The method is essentially an efficient implementation of both Procedure Search and Procedure Corner Points (Figure 4.1).

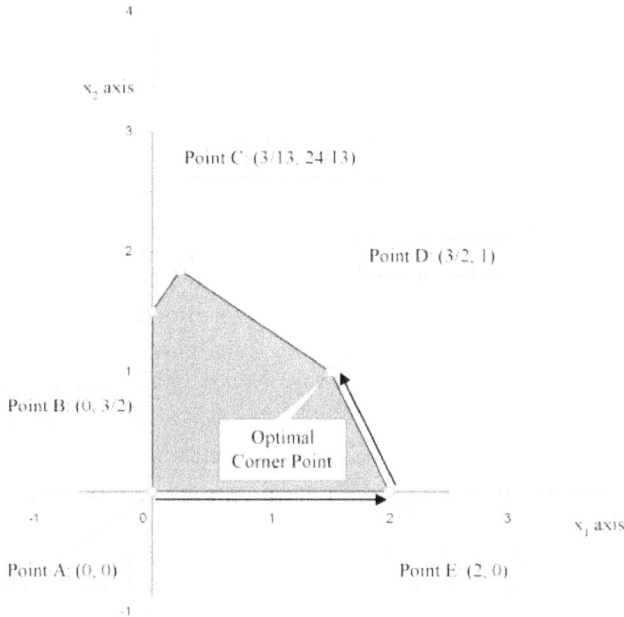

Figure 4.1: Simplex Method Graphically.

For solving, LP problems, first it needs slack variables and setting of Simplex Tableau. It is noted that the Algebraic Method provides all vertices even those which are not feasible. Therefore, it is not an efficient way of solving LP problems with large numbers of constraints. The Simplex Method is a modification of the Algebraic Method, which overcomes this deficiency.

Like Algebraic Method, simplex method is a tabular solution algorithm where each tableau corresponds to movement from one basic variable set BVS to another, making sure that the objective function improves at each iteration till the optimal solution is reached. The following process describes all the steps involved in applying the simplex solution algorithm:

1. Convert the LP to the following form:

    Convert the minimization problem into a maximization one.
    All variables must be non-negative.
    All RHS values must be non-negative (multiply both sides by -1, if needed).
    All constraints must be in £ form (except the non-negativity conditions).
    No strictly equality or constraints are allowed.
    If this condition cannot be satisfied, then use the Initialization of the Simplex Method: Artificial.

2. Convert all £ constraints to equalities by adding a different slack variable for each one of them.

3. Construct the initial simplex tableau with all slack variables in the BVS. The last row in the table contains the coefficient of the objective function (row Cj).

4. Determine whether the current tableau is optimal. That is:
    If all RHS values are non-negative (called, the feasibility condition)
    If all elements of the last row, that is Cj row, are non-positive (called, the optimality condition).
    If the answers to both of these two questions are yes, then stop. The current tableau contains an optimal solution.
    Otherwise, go to the next step.

5. If the current BVS is not optimal, determine, which nonbasic variable should become a basic variable and, which basic variable should become a nonbasic variable. To find the new BVS with the better objective function value, perform the following tasks:

    *   Identify the entering variable: The entering variable is the one with the largest positive Cj value.

    *   Identify the outgoing variable: The outgoing variable is the one with smallest non-negative column ratio. In case of a tie we select the variable that corresponds to the upmost of the tied rows.

    *   Generate new tableau: Do Gauss-Jordan pivoting operation to convert entering column to identity column vector (including the element in the Cj row).

    *   Go to step 4

The simplex method starts at the origin and jumps from a corner point to neighboring corner point till it reaches the optimal corner point. At each simplex iterations, search is done for better solution among that happens within vertices of Simplex. A simplex is n-dimensional space which is the simplest shape with n + 1 vertices. In a triangle which is simplex in 2-dimensional space where a pyramid is simplex in 3-dimensional space, the movements can be seen corresponds to each simplex tableau with specific corner point in graphical method.

Consider the example below which explains about the processing steps involved in simplex method.

Minimize: $-z = -8x_1 - 10x_2 - 7x_3$
St. $x_1 + 3x_2 + 2x_3 \leq 10$
$-x_1 - 5x_2 - x_3 \geq -8$
$x_1, x_2, x_3 \geq 0$

Step 1: Standard Form

Standard form is the base format for a program before it is solved for optimal solution which has following needs:

➤ It should be a maximization problem

➤ The linear constraints should be in less-than-or-equal-to inequality

➤ Variables should be non-negative

The above factors should always be satisfied by transforming given linear program with basic algebra and substitution. Standard form is required as it creates an ideal starting point for solving simplex method as efficiently as compared to other methods. For

$-1 \times (-z = -8x_1 - 10x_2 - 7x_3)$

$z = 8x_1 + 10x_2 + 7x_3$

Maximize: $z = 8x_1 + 10x_2 + 7x_3$

For transforming a minimize linear program model in maximize linear model, multiply both the left and right sides of objective function by -1. Transforming linear constraints from a greater-than-or-equal-to inequality to a less-than-or-equal-to inequality can be done similarly as for an objective function. Also, by multiplying -1 on both sides, inequality can be changed to less-than-or-equal-to.

$-1 \times (-x_1 - 5x_2 - x_3 \geq -8$

$x_1 + 5x_2 + x_3 \leq 8$

As the model is in standard form, slack variables will now be added as explained in step 2.

Step 2: Determine Slack Variables

Slack variables are extra variables in linear constraints of a linear program which

Network Flow Models and Applications
/

are applied to transform from inequality constraints to equality constraints. If model is in standard form, slack variables will always have +1 coefficient, so slack variables are required in constraints to transform into solvable equalities having correct answer.

$8x_1 + 3x_2 + 2x_3 + s1 = 10$

$x_1 + 5x_2 + x_3 + s2 = 8$

$x_1, x_2, x_3, s1, s2 \geq 0$

Once the slack variables are introduced, tableau can be set up for checking optimality as shown in step 3.

Step 3: Setting up the Tableau

Simplex tableau is applied to perform row operations on linear programming model and for checking solution for optimality. The tableau carries coefficient corresponds to linear constraint variables and coefficients of objective function. In tableau as shown, top row shows tableau states in column. The other two rows shows linear constraint variable coefficients from linear programming model while last row shows objective function variable coefficients.

Maximize : $z_1 = 8x_1 + 10x_2 + 7x_3$

s.t. : $x_1 + 3x_2 + 2x_3 + s_1 = 10$

$x_1 + 5x_2 + x_3 + s_2 = 8$

x1	x2	x3	s1	s2	z	b
1	3	2	1	0	0	10
1	5	1	0	1	0	8
-8	-10	-7	0	0	1	0

After completion of tableau, model can be checked for optimal solution as shown in step 4.

Step 4: Check Optimality

It is noted that optimal solution of maximize linear programming model are values that are assigned to variables in objective function for having maximum zeta value. The optimal solution exist on the corner points of the graph of the model. For checking optimality with tableau, values in the last row should be more than or equal to zero. If value is less than zero, then it shows that the variable has not reached its optimal value. In case of earlier tableau, three negative values exists in bottom row that shows that solution is not optimal. If tableau is not optimal, then approach to step 5 which locates pivot variable to base a new tableau on.

Step 5: Identify Pivot Variable

The pivot variable is applied in row operations to find which variable will become the unit value and serves as key factor in conversion of unit value. The pivot variable can be find by looking at the bottom row of tableau and indicator. If the solution is not optimal, then select the lowest negative value in bottom row. Here, a value lying in column of this value will be pivot variable. For finding indicator, divide beta values of linear constraints by corresponding values from column having possible pivot variable. The intersection of row with smallest non-negative indicator and smallest negative value in bottom row will act as pivot variable.

x1	x2	x3	s1	s2	z	b	Indicator
1	3	2	1	0	0	10	10/3
1	(5)	1	0	1	0	8	8/5
-8	-10	-7	0	0	1	0	

Smallest Value

As the new pivot variable is obtained, then new tableau can be created in step 6 so as to optimize the variable and located for new possible optimal solution.

Step 6: Create the New Tableau

The new tableau is applied to find new possible optimal solution. As the pivot variable is obtained in step 5, then row operations can be done so as to optimize pivot variable by keeping the rest of tableau same.

To optimize pivot variable, transform it in unit value which can be done by multiplying row with pivot variable by reciprocal of pivot value. Here it is shown that the pivot variable is originally 5, which on so multiplying with will become:

x1	x2	x3	s1	s2	z	b	
1/5	(1)	1/5	0	1/5	0	8/5	Pivot row

Once the unit value is obtained, other values in column having unit value will become zero as x_2 in second constraint is optimized that needs x_2 in other equations as zero.

x1	x2	x3	s1	s2	z	b	
	0						
1/5	(1)	1/5	0	1/5	0	8/5	Pivot row
	0						

Pivot Column

For having similar tableau, other variables not having in pivot column or pivot row should be calculated by new pivot values. For each new value, multiply negative of value in old pivot column by value in new pivot row which corresponds to value to be calculated and add it to old value from old tableau to have new value for new tableau which is shown in next step.

New tableau value = (Negative value in old tableau pivot column) x (value in new tableau pivot row) + (Old tableau value)

x1	x2	x3	s1	s2	z	b
1	3	2	1	0	0	10
1	(5)	1	0	1	0	8
-8	-10	-7	0	0	1	0

Old pivot column

Old Tableau:

New Tableau:

x1	x2	x3	s1	s2	z	b
2/5	0	7/5	1	-3/5	0	26/5
1/5	(1)	1/5	0	1/5	0	8/5
-6	0	-5	0	2	1	16

New pivot row

Once the new tableau has been done, check the model for optimal solution as shown in step 7.

Step 7: Check Optimality

As described in step 4, optimal solution of maximize linear program model are values that are given to variables as objective function for maximum zeta value. Optimality requires checking after every new tableau for identifying new pivot variable. It is observed that a solution is optimal if values in bottom row are more than or equal to zero. If values are greater than or equal to zero, then solution is optimal and then there will be no need from steps 8 to 11, but if the values are negative, then the solution will not be optimal and there will a need for new pivot point which can be shown in step 8.

Step 8: Identify New Pivot Variable

If the solution is not optimal, a new pivot variable is required. As the pivot variable is introduced in step 5 used in row operations for finding unit value of variable which serves as main factor in conversion of unit value. The pivot variable is recognized by intersection of row with lowest non-negative indicator and lowest negative value in bottom row.

x1	x2	x3	s1	s2	z	b	Indicator
2/5	0	7/5	1	-3/5	0	26/5	(26/5) / (2/5) = 13
1/5	1	1	0	1/5	0	8/5	(8/5) / (1/5) = 8
-6	0	-5	0	2	1	0	

Smallest Value

With the identification of new pivot variable, new tableau can be obtained as explained in step 9.

Step 9: Create New Tableau

Once the new pivot variable is obtained, a new tableau is required. From step 6, tableau is used to optimize pivot variable by keeping remaining tableau same.

Make pivot variable 1 by multiplying the row having pivot variable by reciprocal of pivot value. In tableau as shown below, pivot value was, where everything is multiplied by 5.

x1	x2	x3	s1	s2	z	b
1	5	1	0	1	0	8

Also, make the other values in column of pivot variable zero by taking negative of old value in pivot column and multiplying it by new value in pivot row which is further added to old value which is being replaced.

x1	x2	x3	s1	s2	z	b
0	-2	1	1	-1	0	2
1	5	1	0	1	0	8
0	30	1	0	8	1	64

Step 10: Check Optimality

Optimality can be checked by considering new tableau. From step 4, optimal solution exists when all values in bottom row are more than or equal to zero. If all values are more than or equal to zero, then skip to step 12 since optimality is achieved while if negative values still exist, in such case, repeat steps 8 and 9 till an optimal solution is obtained.

Step 11: Identify Optimal Values

As the tableau results optimal, then optimal values can be traced out by distinguishing basic and non-basic variables. Basic variable have single 1 value in its column while rest values will be zeros. If a variable is not meeting such criteria, then it is considered as non-basic as the optimal solution of that variable is zero. When the variable is basic, then the row has 1 value that corresponds to beta value which shows optimal solution for given variable.

x1	x2	x3	s1	s2	z	b
0	-2	1	1	-1	0	2
1	5	1	0	1	0	8
0	30	1	0	8	1	64

Basic variables: x_1, s_1, z

Non-basic variables: x_2, x_3, s_2

For variable x_1, 1 is found in second row which shows that optimal x_1 value is found in second row of beta values, which is 8.

Variable s_1 has 1 value in first row that shows optimal value as 2 from beta column. Due to s_1 being slack variable, it is not actually included in optimal solution as variable is not contained in the objective function.

The zeta variable has value 1 in last row which shows maximum objective value as 64 from beta column.

The final solution shows each of variables with value:

$x_1 = 8$ $s_1 = 2$

$x_2 = 0$ $s_2 = 0$

$x_3 = 0$ z = 64

Hence the maximum optimal value is 64 which is found at (8,0,0) of objective function.

Example

Maximize 5X1 + 3X2

Subject to:

2X1 + X2 = 40

X1 + 2X2 = 50

and both X1, X2 are non-negative.

After adding two slack variables S1 and S2 the problem is equivalent to:

Maximize $5X_1 + 3X_2$

Subject to:

$2X_1 + X_2 + S_1 = 40$

$X_1 + 2X_2 + S_2 = 50$

and variables X1, X2, S1, S2 are all non-negative.

The initial tableau is:

BVS	X1	X2	S1	S2	RHS	Column Ratio (C/R)
S1	[2]	1	1	0	40	40/2
S2	1	2	0	1	50	50/1
Cj	5	3	0	0		

The solution shown by this tableau is: S1 = 40, S2 = 50, X1 = 0, and X2 = 0. This solution is the origin, shown in our graphical method.

This table is not optimal since some of Cj elements are positive. The incoming variable is X1 and the outgoing variable is S1 (by C/R test). The pivot element is in the bracket. After pivoting, we have:

BVS	X1	X2	S1	S2	RHS	Column Ratio (C/R)
X1	1	1/2	1/2	0	20	20/(1/2)=40
S2	0	[3/2]	-1/2	1	30	30/(3/2)=10
Cj	0	1/2	-5/2	0		

The solution to this tableau is: X1 = 20, S2 = 30, S1 = 0, and X2 = 0. This solution is the corner point (20, 0), shown in our graphical method.

This table is not optimal, since some of Cj elements is positive. The incoming variable is X2 and the outgoing variable is S2 (by C/R test). The pivot element is in the bracket. After pivoting, we have:

BVS	X1	X2	S1	S2	RHS
X1	1	0	2/3	-1/3	10
X2	0	1	-1/3	2/3	20
Cj	0	0	-7/3	-1/3	

The solution to this tableau is: X1 = 10, X2 = 20, S1 = 0, and S2 = 0. This solution is the corner point (10, 20), shown in our graphical method.

This tableau is optimal, since all Cj elements are non-positive and all RHS are non-negative. The optimal solution is X1 = 10, X2 = 20, S1 =0, S2 = 0. To find the optimal value, plug in this solution into the objective function 5X1 + 3X2 = 5(10) + 3(20) = $110.

4.2 Network Flow Model

Consider a model representation:

$$Min \sum_{(i,j)\in A} c_{ij}x_{ij}$$

$$\sum_{i:(i,k)\in A} x_{ik} - \sum_{j:(k,j)\in A} x_{kj} = b_k \quad k \in N$$

From the above model, AX=b, we have:

Matrix A is the node-arc incidence matrix of the network and is a n by m matrix

Matrix A is singular

There is a spanning tree of the undirected network graph corresponding to a basic solution

On removing any one redundant equation, a new system and a new matrix obtained as A: $\tilde{A}X = \tilde{b}$.

Now,

The new A matrix corresponding a

The new matrix A corresponds to a cycle, which has linearly dependent columns

(n-1) columns of new A that forms a non-singular matrix ((n-1) by (n-1)) corresponds to a spanning tree

4.2.1 Finding in Initial Feasible Solution

By using a spanning tree we can solve a system of equations to get an initial solution. For example consider a spanning tree:

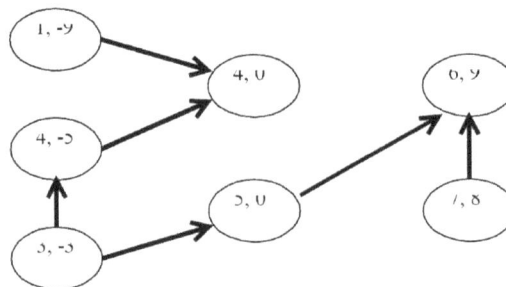

Spanning Tree.

Here, first number of two numbers in an ellipse shows that the number of nodes with second shows as demand of this node. Now the corresponding system of equations:

$$-x_{14} = -9$$
$$x_{14} + x_{24} = 0$$
$$x_{32} - x_{24} = -5$$
$$-x_{35} - x_{32} = -3$$
$$x_{35} - x_{56} = 0$$
$$x_{56} + x_{76} = 9$$
$$-x_{76} = 8$$

The easy way to solve this system is to first find equation with only one variable; that is finding a leaf in the corresponding spanning tree. In this, the solution is not feasible as it violates the nonnegative constraints.

Further, just like phase 1 in simplex method a procedure is required to get feasible solution which can be described.

Finding node receiving from supplying nodes and sending it to all other non-supply nodes, but if this is not, then qualified node cannot be obtained, in such case, selection of node and construction of arcs requires sending from supplying nodes to particular node or sending from this node to other non supply nodes.

Further on solving the problem as per the spanning tree, if such problem does not have zero objective value, then original problem is infeasible, and if it is, in such case a feasible solution is obtained.

4.2.2 Using dual Cost Tree to find Optimal Solution

Considering dual problem of minimum network flow problem as:

$$Max \quad \sum_{k \in N} b_k y_k$$

$$y_j - y_i \le c_{ij} \quad (i, j) \in A$$

In the above system, it is known that primal problem has n-1 non basic v. If we have an initial primal flow solution:

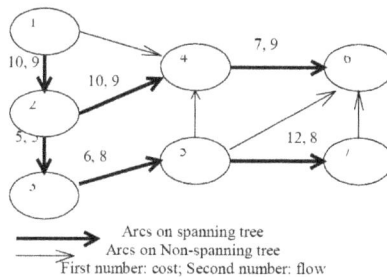

Initial Primal Flow Model.

The dual cost solution:

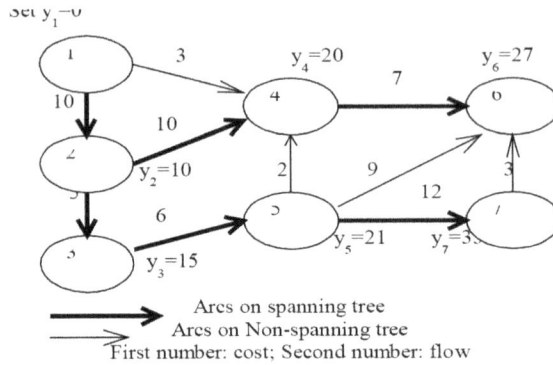

Set $y_1=0$

$y_4=20$ $y_6=27$

Arcs on spanning tree
Arcs on Non-spanning tree
First number: cost; Second number: flow

Now checking optimality of primal problem:

Here, arc 14 does not satisfy optimality, so this arc should enter into the spanning tree, which corresponds to non-basic variable. On adding arc 14 in primal flow tree, a cycle is obtained where an arc is dropped to retain a spanning tree as shown in primal flow tree:

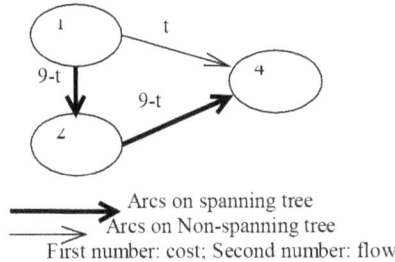

Arcs on spanning tree
Arcs on Non-spanning tree
First number: cost; Second number: flow

Primal Flow Tree.

From the above tree, set t=9 and the whole primal flow tree will result as:

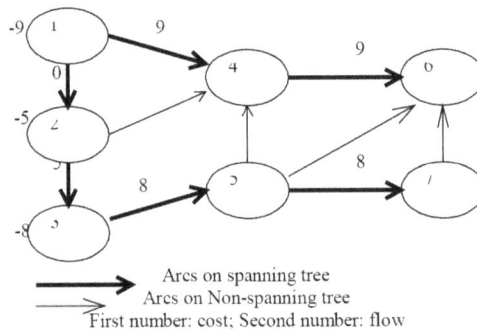

Arcs on spanning tree
Arcs on Non-spanning tree
First number: cost; Second number: flow

Spanning Tree.

Dual cost tree:

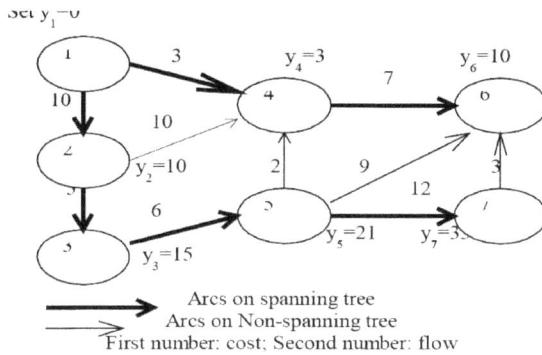

Set $y_1=0$

Arcs on spanning tree
Arcs on Non-spanning tree
First number: cost; Second number: flow

Here the optimality is checked using arcs that are off the spanning tree which satisfies the inequalities resulting in optimal solution.

4.2.3. Shortest Path Problem

Now we will look for shortest paths from a source to all other nodes in a network.

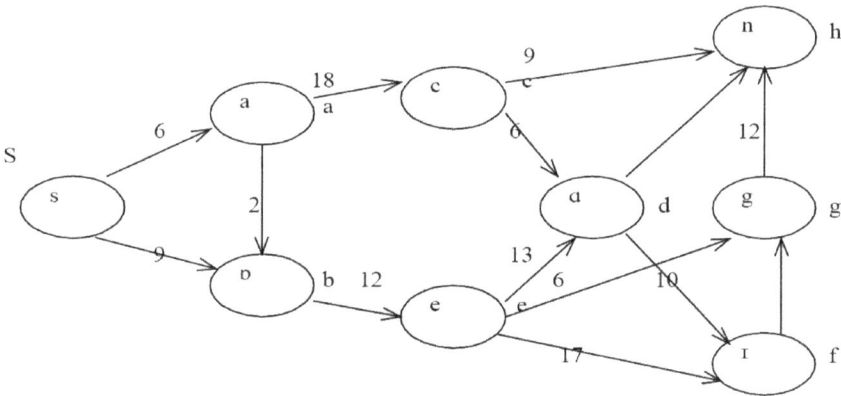

For this, label the notes which starts from the source node (node, distance) and consider the following steps:

Step 1: label source node:

S

S, 0

Step 2: separate labeled nodes from unlabeled nodes. And label node that has the shortest distance from separated labeled notes set.

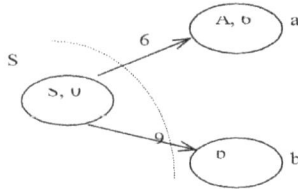

Keep doing step2 until all nodes have been labeled.

Step 3:

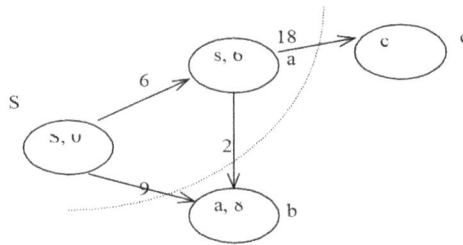

Step 4:

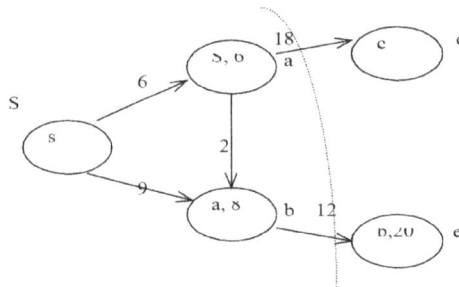

Step 5:

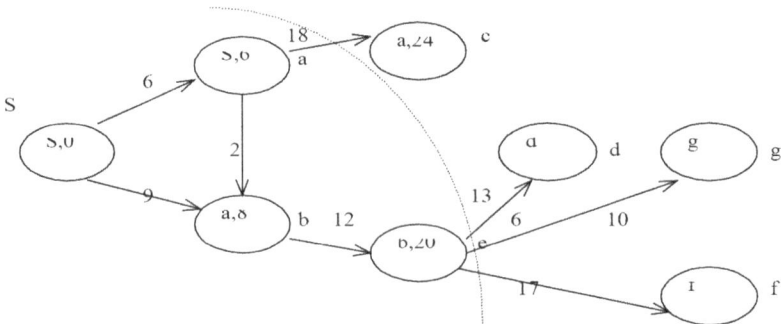

Step 6:

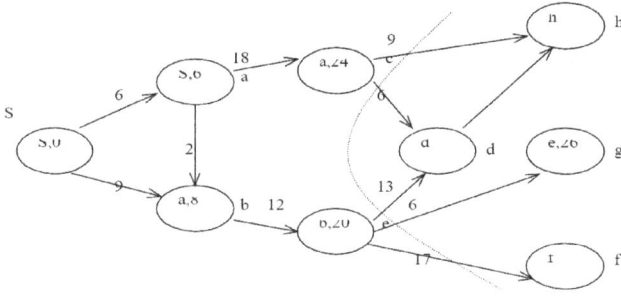

Step 7:

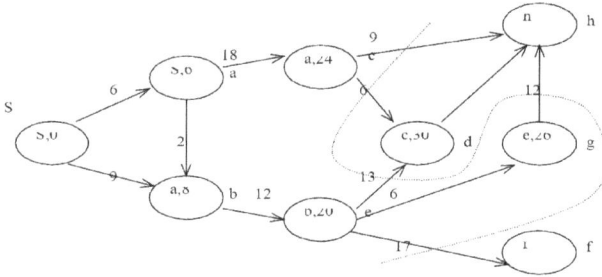

Step 8:

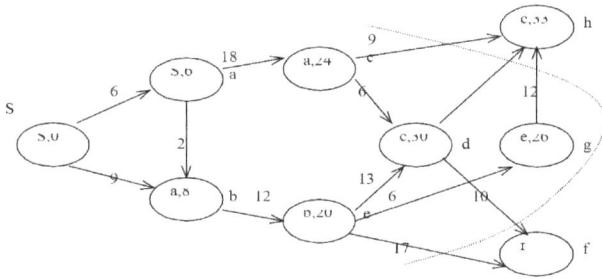

Step 9:

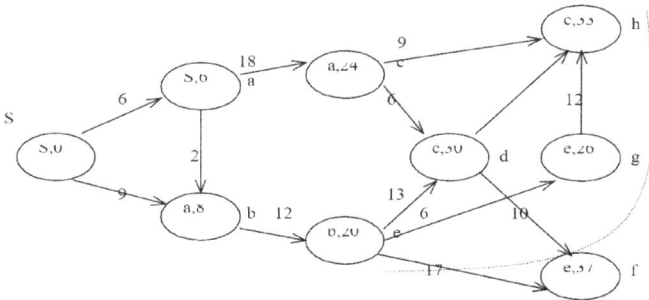

From these, we get the solution as:

(s, a)=6

(s, b)=8

(s, c)=24

(s, d)=30

(s, e)=20

(s, f)=37

(s, g)=26

(s, h)=33

4.2.4 Maximum Flow Problem

Considering maximum flow problem, where a network that supplies node and demand node while others are intermediate nodes. The idea behind it is to push a flow from supplying node to demand node. The supply of flow depends on the capacities of arcs that are connected among the nodes which are not infinite (Figure 4.2)

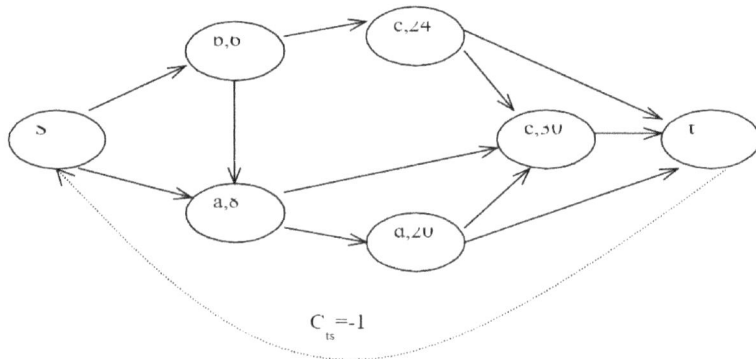

Figure 4.2: Network Flow.

For this, use a dummy arc from demanding node t to supplying node s and assume c_{ts} as negative . At same time, let all c_{ij}=0 and all demands of all nodes to zero (b_l=0). As the only nonzero cost of network is negative, so on transferring maximum flow from demanding node t to supplying node s, for maximum profit, hence we have:

$$Max \quad - x_{ts}$$
$$\sum_{i:(i,k)\in A} x_i - \sum_{j:(k,j)\in A} x_j = 0$$
$$0 \leq x_j \leq u_j$$

4.3 General Minimal Cost Flow Problem

Minimum cost flow problem is an optimization and decision problem which finds the cheapest possible way of sending certain amount of flow through a flow network. A typical application of this problem involves finding the best delivery route from a factory to a warehouse where the road network has some capacity and cost associated. The minimum cost flow problem is one of the most fundamental among all flow and circulation problems because most other such problems can be cast as a minimum cost flow problem and also that it can be solved very efficiently using the network simplex algorithm.

The minimum cost flow problem finds a feasible flow of minimum total cost from a set of supply nodes to a set of demand nodes in a network with capacity constraints (lower and upper bounds) and arc costs.

If $G = (V, A)$ be a digraph, *lower* $: A \rightarrow R$ *upper* $: A \rightarrow R \cup \{+\infty\}$, shows lower and upper bounds for the flow values on arcs, for which *lower (uv)* $\leq$ *upper (uv)* must hold for all $uv \in A$, *cost* $: A \rightarrow R$ which shows cost per unit flow on arcs and $sup : V \rightarrow R$ denotes the signed supply values of the nodes.

If , *sup (u)* > 0, then u is supply node with *sup(u)* supply, if *sup (u)* < 0, then u is a demand node with *−sup(u)* demand. A minimum cost flow is an $f : A \rightarrow R$ solution of the following optimization problem.

$$\min \sum_{uv \in A} f(uv). \, \text{cost}(uv)$$

$$\sum_{uv \in A} f(uv) - \sum_{vu \in A} f(vu) \geq \sup(u) \quad \forall u \in V$$

$$lower(uv) \leq f(uv) \leq upper(uv) \quad \forall uv \in A$$

Here, the sum of supply values, *i.e.* $\sum_{u \in V} \sup(u)$ must be zero or negative in order to have a feasible solution (since the sum of the expressions on the left-hand side of the inequalities is zero). It means that the total demand must be greater or equal to the total supply and all the supplies have to be carried out from the supply nodes, but there could be demands that are not satisfied. If $\sum_{u \in V} \sup(u)$ is zero, then all the supply/demand constraints have to be satisfied with equality, *i.e.* all demands have to be satisfied and all supplies have to be used.

The variation of this problem is to find a flow which is maximum, but has the lowest cost among the maximum flow solutions which could be minimum-cost maximum-flow problem which is useful for finding minimum cost maximum matching's.

With some solutions, finding the minimum cost maximum flow instead is straightforward. If not, one can find the maximum flow by performing a binary search on d. A related problem is the minimum cost circulation problem, which can be used for solving minimum cost flow. This is achieved by setting the lower bound on all edges to zero, and then making an extra edge from the sink t to the source s, with capacity c(t,s)=d and lower bound l(t,s)=d, forcing the total flow from s to t to also be d.

The problem can be specialized into two other problems:

> ➤ if the capacity constraint is removed, the problem is reduced to the shortest path problem,

> ➤ if the costs are all set equal to zero, the problem is reduced to the maximum flow problem.

Further related to max flow problem is minimum cost flow problem where each arc in the graph has a unit cost for transporting material across it. The problem is to find a flow with the least total cost. The min cost flow problem also has special nodes, called supply nodes or demand nodes, which are similar to the source and sink in the max flow problem. Material is transported from supply nodes to demand nodes.

> ➤ At a supply node, a positive amount — the supply — is added to the flow. A supply could represent production at that node, for example.

> ➤ At a demand node, a negative amount — the demand — is taken away from the flow. A demand could represent consumption at that node, for example.

> ➤ For convenience, we'll assume that all nodes, other than supply or demand nodes, have zero supply.

For the min cost flow problem, we have the following flow conservation rule, which takes the supplies and demands into account:

4.4 Minimal Cost Calculation

Use of breakeven and minimum-cost-point formulas require collection of unit costs. Unit costs can be divided into subunits, each of which measures the cost of a certain part of the total. A typical unit cost formula might be $X = a + b + c$. In this, X is cost per unit volume and subunits a, b, c are distance, volume, area or weight.

Suppose the cost of harvesting from felling to loading on trucks is studied. If X is cost per cubic meter of wood loaded on the truck, we could represent the total cost per unit as $X = A + B + Q + L$, where

A = cost per unit of felling

B = cost of bucking

Q = cost of skidding

L = cost of loading

To find cost per subunit for felling, bucking, skidding, and loading, the factors which determine production and cost must be specified. Functional forms for production in road construction and harvesting are shown. Examples for felling and skidding describes that for felling, tree diameter may be an important explanatory variable. For a given felling method, the time required to fell the tree might be expressed as $T = a + b\ D^2$ where T is time to fell the tree, b is felling time required per cm of diameter, D is the tree diameter and "a" represents the felling time not explained by tree diameter-such as for walking between trees.

The production rate is equal to the tree volume divided by the time per tree. The unit cost of felling is equal to the cost per hour of the felling operation divided by the hourly production or $A = C/P = C/(V/T) = C\ (a + B\ D^2)/V$ where C is cost per hour for felling method, P is production per hour, V is volume per tree and T is time per tree.

Example:

Finding felling unit cost for 60 cm tree if cost per hour of a man with power saw is 5.00, the tree volume is 3 cubic meters, and the time to fell the tree is 3 minutes plus 0.005 times the square of the diameter.

$T = 3 + .005\ (60)\ (60)$

$= 21$ min

$= .35$ hr

$P = V/T$

$= 3.0/.35$

$= 8.57$ m^3/hr

$A = C/P$

$= 5.00/8.57$

$= 0.58/$m^3

In skidding, for example, if logs were being skidded directly to a road, then the distance skidded is an important factor and the stump to truck unit cost might be written as

$X = A + B + Q + L$

$X = A + B + F + C(D/2) + L$

where the skidding subunit Q has been replaced by symbol F representing fixed costs of skidding such as hooking, unhooking and decking and C(D/2) represents that part of the skidding cost that varies with distance. C is the cost of skidding a unit distance such as one meter and D/2 represents the average skidding distance in similar units.

It is important to note that the average skidding cost occurs at the average skidding distance only when the skidding cost, C does not vary with distance. If C varies with distance, as for example, with animal skidding where the animal can become increasingly tired with distance, the average skidding cost does not occur at the average skidding distance and substantial errors in unit cost calculations can occur if the average skidding distance is used.

If logs were being skidded to a series of secondary roads running into a primary road, then the expression C(D/2) would be replaced by the expression C(S/4) and the cost of truck haul on the secondary roads would appear as a separate item. In the expression C(S/4), the symbol S represents the spacing of the secondary roads and the distance S/4 is the average skidding distance if skidding could take place in both directions. Therefore, the expression C(S/4) would define the variable skidding cost in terms of spacing of the secondary roads.

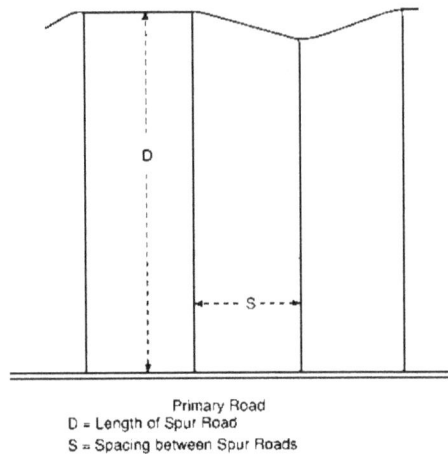

Primary Road
D = Length of Spur Road
S = Spacing between Spur Roads

Figure 4.3: Way Skidding to Continuous Landings among Spur Roads.

A formula for the cost of logs on trucks at the primary road under these circumstances would be

$$X = A + B + F + C(S/4) + L + H(D/2)$$

where D/2 is the average hauling distance along the secondary road and H is the variable cost of hauling per unit distance. The formula can be extended still further to include the cost of the secondary road system by defining the road construction cost per meter R, and the volume per square meter, V. Then, the formula becomes

$$X = A + B + F + C(S/4) + L + H(D/2) + R/(VS)$$

4.4.1 Cost Equations

In earlier equation, the spacing between skidding roads increases, so skidding unit cost also increase, while road unit costs decrease. With total cost equation, the cost tradeoffs among skidding distance and road spacing is seen where formula for road spacing which minimizes costs will result as:

$dX/dS = C/4 - R/(VS^2) = 0$

or

$S = (4R/CV)^{.5}$

An alternative method is to compare total costs for various road spacings. The total cost method has become less laborious with the use of programmable calculators and microcomputers. It provides information on the sensitivity of total unit cost to road spacing without having to evaluate the derivative of the cost function.

Example

Given the following table of unit costs, what is the effect of alternative spur road spacings on the total cost of wood delivered to the main road if 50 m^3 per hectare is being cut and the average length of the spur road is 2 km. The cost of spur roads includes landings.

Activity	Unit	Cost
Fell	/m^3	0.50
Buck	/m^3	0.20
Skid	/m^3	2.00 (fixed cost)
Skid	/m^3-km	2.50 (variable cost)
Load	/m^3	0.80
Transport	/m^3-km	0.15
Roads	/km	2000

Since only the skidding costs and spur road costs are affected by the road spacing, the total unit cost can be expressed as

$X = A + B + F + C(S/4) + L + H(D/2) + R/(VS)$
$X + 0.50 + 0.20 + 2.00 + C(S/4) + 0.80 + .15 (1) + R/(VS)$
$X = 3.65 + C(S/4) + R/(VS)$

To find different road spacings, vary spur road spacing S and calculate total unit costs. It is important to use dimensionally consistent units, if left side of equation is in /m^3, then right side of equation should be in /m^3. It is possible if all volumes, costs and distances are expressed in meters like volume cut/m^2, skidding cost/m^3 / meter and road cost/meter.

For example, the total cost for a spur road spacing of 200 meters is 3.65 + (2.5/1000) (200/4) + (2000/1000)/[(50/10000) (200)] or 5.78 per m^3.

Spur Road Spacing, m	Total Unit Cost, /m^3
200	5.78
400	4.90
600	4.69
800	4.65
1000	4.68
1200	4.73
1400	4.81
1600	4.90
1800	5.00
2000	5.10

The road spacing which minimized total cost could be interpolated from the table or calculated from the formula

$S = (4R/CV)^{.5}$

$$S = \frac{[4 \times 2000 / 1000]^{5}}{(2.5 / 1000)(50 / 10000)}$$

S = 800 m.

When costs have been collected in a form which permits unit costs to be developed from them, not only is it possible to predict costs, it is also possible to adjust conditions so that minimum cost can be achieved. Too often, recorded costs are only "experience figures". They are usually made available in a form which can be used to predict costs only under conditions that closely conform to those existing where and when the recorded costs were collected. This is not true of unit costs, which can be fitted into the framework of many different harvesting situations and can be made to tell the story of the future as well as that of the past.

A wide range of cost control formulas can be derived. Typical problems include:

1. The economic location of roads and landings. - The calculation of the optimal spacing between spur roads and landings subject to one-way skidding, two-way skidding, skidding on slopes, linear and nonlinear skidding cost functions.

2. The economic service standard for roads. - The comparison of the benefits of lower haul costs and road maintenance costs as a function of increased initial investment. The calculation of the optimal length of swing roads as a function of the tributary volume.

3. The economic selection of equipment for road systems fixed by topography or other factors. - The identification of the breakeven points between alternative skidding methods which have different fixed and variable operating costs.

4. The economic spacing of roads which will be served by two types of skidding machines. - For example, machines used to skid saw timber and to relog for fuel wood.

5. The economic spacing of roads which will be reused in future time periods.

Another important application of unit costs is in choosing between alternative harvesting systems.

Example

A forest manager is developing an area and is trying to decide between harvesting methods. He has two choices of skidding systems, two choices of road standards, and two choices of trucks. If larger skidding equipment is selected to bring the logs to the landing, he can still choose to buck them into smaller logs on the landing. We assume that bucking on the landing will not affect log quality or yield.

The managers staff has developed the relevant unit costs, which are summarized in Tables.

	Small Equipment $/m^3$	Large Equipment $/m^3$
Fall, buck	0.70	0.50
Skid	1.70	2.55
Load	1.00	0.80
Transport	1/	1/
Unload	0.40	0.30
Process	-	0.05 2/

		Small Equipment $/m^3$	Large Equipment $/m^3$
Road			
	High Standard	1.30	1.30
	Low Standard	1.00	1.00
Transport			
	High Standard	3.50	3.00
	Low Standard	4.00	3.40

These choices can be viewed as a network. Now verify that the least cost path is obtained by using larger skidding equipment and trucks and constructing the higher standard road. The total unit cost will be 8.50 per m³. A important point is the ease at which these problems can be analyzed, once the unit costs have been derived. In turn, the derivation of the unit costs is facilitated by having machine rates available.

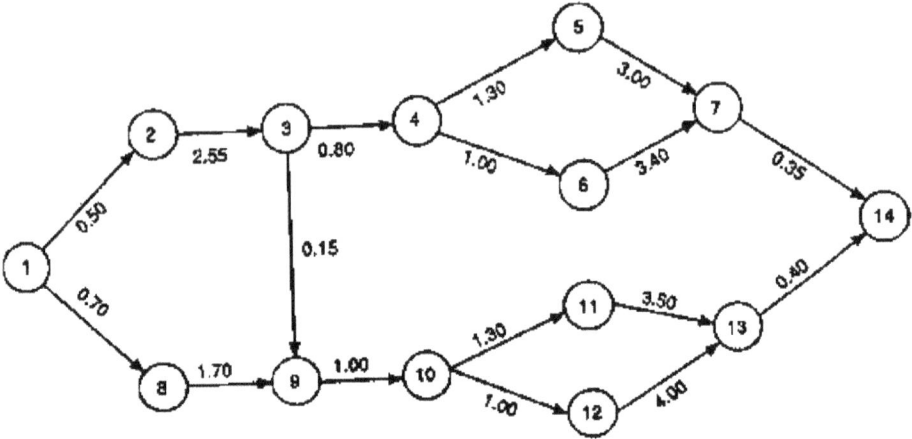

Figure 4.4: Network Diagram for Equipment Choice.

4.5 Network Simplex Algorithm Implementation

Network simplex algorithm is a graph theoretic specialization of simplex algorithm which is usually formulated in terms of standard problem, minimum-cost flow problem and can be efficiently solved in polynomial time. The network simplex method works very well in practice, typically 200 to 300 times faster than the simplex method applied to general linear program of same dimensions.

Network simplex method is adaptation of bounded variable primal simplex algorithm which is shown as rooted spanning tree of underlying network where variables are shown by arcs and simplex multipliers by node potentials. At each iteration, an entering variable is selected by some pricing strategy, based on the dual multipliers and forms a cycle with the arcs of the tree. The leaving variable is the arc of the cycle with the least augmenting flow. The substitution of entering for leaving arc, and the reconstruction of the tree is called a pivot. When no non-basic arc remains eligible to enter, the optimal solution has been reached.

Let us use the example

Maximise $z = \chi + y$

When $3\chi + 4y + u = 12$

                 $3\chi + 2y + v = 9$

With $\chi \geq 0$, $y \geq 0$, $u \geq 0$ and $v \geq 0$

Firstly we need to rewrite the objective equation = 0 by rearranging it.

$$z - \chi - y = 0$$

Here it is required to rearranged $\chi + y - z = 0$ but for Simplex Method to work, xalues of χ and y are required to be negative at start. Then writing the equations in tableau as shown:

χ	y	u	v	z	
3	4	1	0	0	12
3	2	0	1	0	9
-1	-1	0	0	1	0

The first two rows represent the constraint equations with the slack variables and the bottom row represents the objective function. An iteration of the simplex method moves us along a line of the boundary of the feasible region to another vertex.

It is observed that either it is required to increase χ or y that will move along χ or y axes respectively. It is required that the variable which are always changed first will be maximum highest entry in objective column. The column in which such variable lies is pivot column.

In the example, χ and y will have both values of –1 in objective row so, selecting χ

χ	y	u	v	z	
3	4	1	0	0	12
3	2	0	1	0	9
-1	-1	0	0	1	0

Pivot column

On increasing χ, it is clear from below graph that, at point c which is the vertex, a pause is required that serves as the line $3\chi + 2y = 9$.

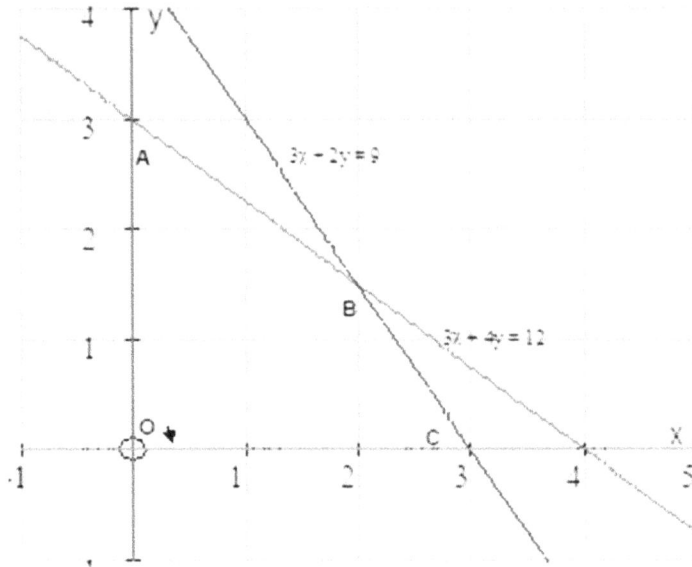

Figure 4.5: Graph

We can find the information from the tableau by dividing the right hand column entries (12 and 9) by χ column entries. The smallest possible values will occur in row corresponds to correct line which is known as pivot row. The χ value in the first row is 3 and $12 \div 3 = 4$; the χ value in the second row is also 3 and $9 \div 3 = 3$. As 3 is smaller than 4 the second row becomes our pivot row.

χ	y	u	v	z		
3	4	1	0	0	12	$12 \div 3 = 4$
3	2	0	1	0	9	$9 \div 3 = 3$ Pivot row
-1	-1	0	0	1	0	

Pivot column

Here it is observed that the value which is in both the pivot column and pivot row is known as pivot element. (box 3). It is required to make the pivot element to have value 1, for which, applying division and perform the similar operations to all numbers in that row. In the example, change 3 to 1, as to divide by 3 throughout in the row.

	χ	y	u	v	z	
	3	4	1	0	0	12
R2 ÷ 3	1	2/3	0	1/3	0	3
	-1	-1	0	0	1	0

Now make all elements in pivot column as zero by carrying out row operations. To change the '3' in row 1 to '0', subtract a multiple of row 2 from it, *i.e.* 3 times row 2.

	χ	y	u	v	z	
R1 - 3R2	0	2	1	-1	0	3
	1	2/3	0	1/3	0	3
	-1	-1	0	0	1	0

Now taking value in row 2 and add it to the value in row 3 to obtain a new value of 'z'.

	χ	y	u	v	z		
	0	2	1	-1	0	3	
	1	2/3	0	1/3	0	3	
R3 + R2	0	-1/3	0	1/3	1	3	New value of z

From the tableau, set y and v to '0' and find χ = 3 and z = 3, hence we get vertex where χ = 3 and y = 0 giving the objective

z = 3

It is a single iteration of the simplex algorithm. Also, still we have negative value in objective row, so repeating process again, and considering tableau as:

χ	y	u	v	z	
0	2	1	-1	0	3
1	2/3	0	1/3	0	3
0	-1/3	0	1/3	1	3

Since the column y has highest negative value in objective row, so it becomes the pivot column and on division, row *R1* will be pivot row which will be 1.5 smaller than 4.5.

χ	y	u	v	z	
0	**2**	1	-1	0	3
1	2/3	0	1/3	0	3
-1	-1	0	1/3	1	3

$3 \div 2 = 1.5$ **Pivot row**

$3 \div 2/3 = 4.5$

Pivot column

From this the pivot element is 2 so we need to divide row 1 by 2 giving

	χ	y	u	v	z	
R1 ÷ 2	0	1	1/2	-1/2	0	3/2
	1	2/3	0	1/3	0	3
	0	-1/3	0	1/3	1	3

We now want to make the y value zero in the other rows, so

$$R2 - \frac{2}{3}R1$$

$$R3 + \frac{1}{3}R1$$

giving

	χ	y	u	v	z	
	0	1	1/2	-1/2	0	3/2
R2 - 2/3R1	1	0	-1/3	2/3	0	2
R3 + 1/3R1	0	0	1/6	1/6	1	3.5

Now set u and v to zero which gives $\chi = 2$, $y = \frac{3}{2}$ and $z = 3\frac{1}{2}$ (3.5) which corresponds to point B on the graph. It is known that, B will be the optimal solution, so an optimal solution is obtained where an objective row will contain no negative values.

Example

Use a simplex tableau to solve the Linear Programming problem. Begin by pivoting on an element chosen from the χ column. Write down the values of χ, y and f at the end of each iteration.

Maximise	$f = 9\chi + 4y$
Subject to	$3\chi + 4y \leq 48$
	$2\chi + y \leq 17$
	$3\chi + y \leq 24$
	$\chi \geq 0, y \geq 0$

Solution

Using slack variables

$3\chi + 4y + u = 48$

$2\chi + y + v = 17$

$3\chi + y + w = 24$

$-9\chi - 4y + f = 0$

The Simplex Tableau will be:

x	y	u	v	w	f	
3	4	1	0	0	0	48
2	1	0	1	0	0	17
3	1	0	0	1	0	24
-9	-4	0	0	0	1	0

χ is the pivot column

$48 \div 3 = 16$ $\qquad$ $17 \div 2 = 8.5$ $\qquad$ $24 \div 3 = 8$

8 is the lowest so *R3* is the pivot row. This means that 3 is the pivot element.

$R3 \div 3$

	x	y	u	v	w	f	
R1	3	4	1	0	0	0	48
R2	2	1	0	1	0	0	17
R3	1	1/3	0	0	1/3	0	8
R4	-9	-4	0	0	0	1	0

Pivot about the χ values *i.e.* make all the χ values in *R1, R2* and *R4* all equal 0.

$R1 - 3R3$ $\qquad$ $R2 - 2R3$ $\qquad$ $R4 + 9R3$

	x	y	u	v	w	f	
R1	0	3	1	0	-1	0	24
R2	0	1/3	0	1	-2/3	0	1
R3	1	1/3	0	0	1/3	0	8
R4	0	-1	0	0	3	1	72

At this point set y and w to 0 giving

$\chi = 8$, y = 0 and f = 72

Since negative value in present in object row, so on repeating the process and doing second iteration with y as pivot column, since it is the largest negative entry, hence

$24 \div 3 = 8$ $1 \div \dfrac{1}{3} = 3$ $8 \div \dfrac{1}{3} = 24$

3 is the lowest so *R2* is the pivot row which shows that $\dfrac{1}{3}$ is pivot element.

$R2 \div \dfrac{1}{3}$

	x	y	u	v	w	f	
R1	0	3	1	0	-1	0	24
R2	0	1	0	3	-2	0	3
R3	1	1/3	0	0	1/3	0	8
R4	0	-1	0	0	3	1	72

$R1 - 3R2$ $R3 - -R2$ $R4 + R2$

	x	y	u	v	w	f	
R1	0	0	1	-9	5	0	15
R2	0	1	0	3	-2	0	3
R3	1	0	0	-1	1	0	7
R4	0	0	0	3	1	1	75

Here every entries in objective row are non-negative, so this is the optimal solution. Now if we set v and w to 0 we get

$\chi = 7$, y = 3 and f = 75

Example

Maximise the objective function $f = -\chi + 8y + z$

Where $\chi + 2y + 9z \leq 10$

$$y + 4z \leq 10$$

$$\chi \geq 0, y \geq 0, z \geq 0$$

Solution

Now,

Objective function $f = -\chi + 8y + z$ $f + \chi - 8y - z = 0$

Slack variables $\chi + 2y + 9z + u = 10$

$$y + 4z + v = 12$$

Simplex tableau

x	y	z	u	v	f	
1	2	9	1	0	0	10
0	1	4	0	1	0	12
1	-8	-1	0	0	1	0

-8 is largest negative value so pivot about y

$10 \div 2 = 5$ $12 \div 1 = 12$

5 is the lowest entry so R1 is the pivot row which shows that 2 is the pivot element.

$R1 \div 2$

	x	y	z	u	v	f	
R1	1/2	1	9/2	1/2	0	0	5
R2	0	1	4	0	1	0	12
R3	1	-8	-1	0	0	1	0

R2 – R1   R3 + 8R1

	x	y	z	u	v	f	
R1	1/2	1	9/2	1/2	0	0	5
R2	-1/2	0	-1/2	-1/2	1	0	7
R3	5	0	35	4	0	1	40

Now set χ, z, and u to 0 which shows y = 5 and f = 40. Since all values in objective row are non-negative, so this is the optimal solution. The maximum value of 'f' = 40 occurs when $\chi = 0$, y = 5 and z = 0.

4.6 Primal Simplex Method

Considering the network problem which can be solved by using primal simplex in two phases, where first phase starts with artificial arcs for each node and phase second starts with the solution and continues to the optimum solution (Figure 4.6).

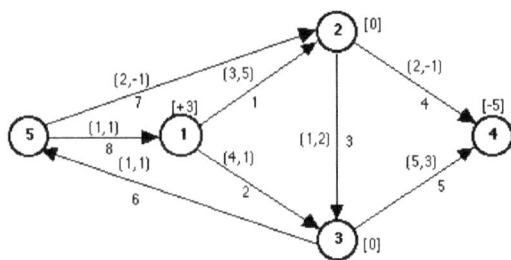

Figure 4.6: Primal Network.

Considering initial solution of artificial arcs which are going to and from slack node that includes for every node in network which serves as initial basis. For each node $i = 1$ to m 1,

a. If node i has $bi > 0$, construct an artificial arc from node i to the slack node. Set the arc flow and the arc capacity to bi.

b. If node i has $bi < 0$, construct an artificial arc from the slack node to node i. Set the arc flow and the arc capacity to $-bi$.

c. For Phase 1, set the cost on the artificial arcs to 1. Set all other arc costs to 0 (Figure 4.7).

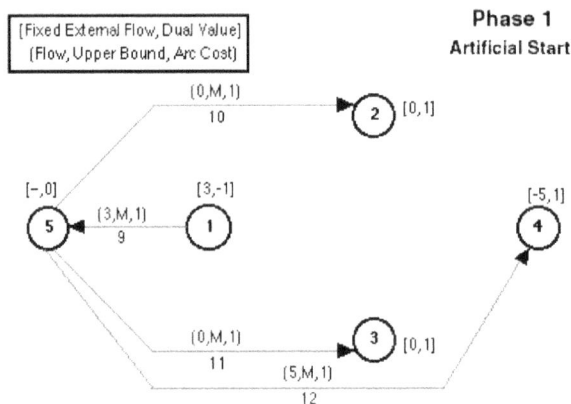

Figure 4.7: Phase 1.

Consider Phase 1:

Assign the arc cost of $+1$ to each of the artificial arcs and a cost of 0 to each of the original arcs.

Using the artificial arcs as the initial basis, solve the network problem with the primal simplex algorithm. If the total cost at optimality is greater than zero, an artificial arc has nonzero flow, and there is no feasible solution to the original problem. Stop and indicate that there is no feasible solution.

If the total cost at optimality is zero, all the artificial arcs have zero flow and a feasible solution has been found. Proceed with phase 2 (Figure 4.8).

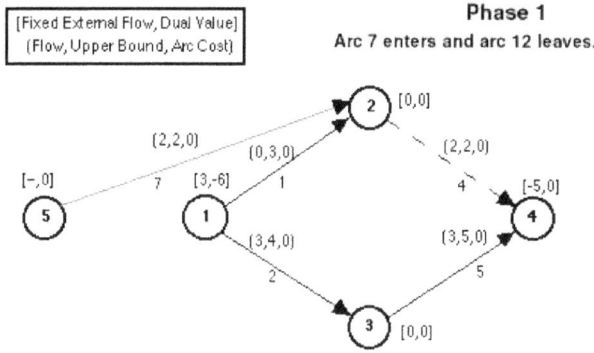

Figure 4.8: Phase 2.

Consider Phase 2:

Assign the original arc costs to the original arcs. Assign 0 cost and 0 capacity to each artificial arc. Starting with the basic solution found in phase 1, solve the network problem with the primal simplex algorithm (Figure 4.9).

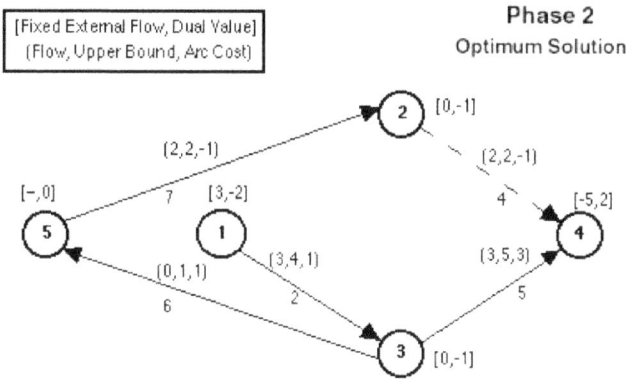

Figure 4.9: Primal Simplex Algorithm.

Now considering primal simplex algorithm for minimum cost network flow programming problem which carries following steps.

Start with a basis, nB, and the sets $n0$ and $n1$, corresponding to nonbasic arcs at their lower and upper bounds respectively, that is feasible for the primal problem. Compute the primal and dual basic solutions for the initial basis.

Compute the reduced costs, dk, for all nonbasic arcs.3. If for each nonbasic arc one of the conditions below holds then *stop* with the optimum solution.

Conditions for Optimality

If $x\text{k} = 0$, then $d\text{k} > 0$.

If $x\text{k} = u\text{k}$, then $d\text{k} < 0$.

Otherwise, select some nonbasic arc that violates the optimality condition and call it the entering arc. Now find the increase or decrease in entering arc that will either drive it to its opposite bound or drive some basic arc to one of its bounds. If the entering arc is driven to its bound. If a basic arc is driven to one of its bounds, let this be the leaving arc. Further change the basis by removing the leaving arc from the basis and adding the entering arc. Compute the primal and dual basic solutions associated with the new basis.

5 | Applications of Network Flow

5.1 Introduction

The applications of network flow will involve physical situations that can be set of water flowing pipes, traffic in network etc in which translation of input data in required graph is intuitive. In many applications of network flow, problems pertaining will not involve physical movement of items through networks. The main type of application of network flow is bipartite matching which involves network flow along few extra sample problems that can be solved with network flow.

Normally it is analysed that flow network has directed graph with nodes and edges that links these nodes serving as designated source and sink node as part of network where every edge has a capacity having maximum amount of flow on particular edge. Further, every node in the network except the source and sink must satisfy the condition that all of the flow coming into the node has to equal all of the flow coming out of the node. The source is intuitively where all flow originates from, and the sink is where all flow must finally pass.

From above, concept of maximum network flow is maximum amount of flow which leaves the source and gives way to sink through network. There are also good strong polynomial time algorithms that makes these problem to solve efficiently which can be surely be applied to many concepts.

5.1.1 Maximum Flow and Minimum Cut

The max-flow min-cut is an important theorem which results in graph theory that shows that a weight of minimum s-t cut in a graph will be equal to the value of maximum flow in a corresponding flow network. As a consequence, every max flow algorithm may be employed to solve the minimum s-t cut problem, and vice versa. To every weighted s-t graph (V, E, s, t, w) we can associate a corresponding flow network (V, s, t, c) in which c(u,v) = w(u,v) if (u,v) ε E, and c(u,v) = 0 otherwise. That is, the edges of the flow network are the same as the edges of the weighted s-t graph, with weights replaced by capacities.

Let f be a maximum flow in a flow network (V, s, t, c). Consider the set S ⊂ V consisting of vertices reachable from S in the residual network corresponding to f. A minimum s-t cut of the corresponding weighted s-t graph (V, E, s, t, w) is then given by $C = \{e = (u,v)\,|\,e \in E, u \in S, u \notin S\}$ Furthermore, $\sum_{v \in V} f(s,v) = \sum_{e \in C} w(e)$, that is, the max flow and the cut thus obtained are equal in value.

The minimum s-t cut problem can be solved by converting all edge weights to capacities and then finding a maximum flow in resulting flow network. To have the weight of min cut, find value of the max flow while to find cut explicitly, it is right to do using residual graph corresponds to max flow. Here "s part" of the s-t cut consist of vertices reachable from s in residual graph, with the remainder "t part". The edges to be cut are from "s part" to "t part".

Example 5.1

In a network with four edges, the source is on top of network and sink is below the network. Each edge has a maximum flow of 3. Find the flow which can pass through this network at any given time?

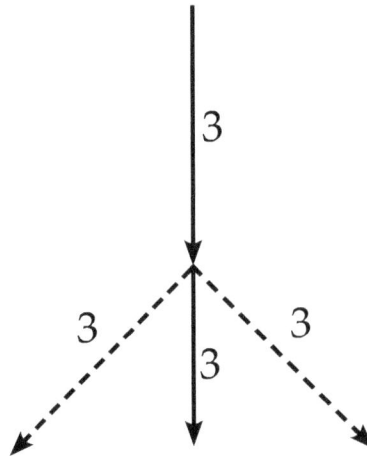

It is observed from network that the flow will be 3 as the bottom three edges can pass 9 between the three of them and the limiting factor is at the top edge that pass 3 at a time. As it is an intuition behind max-flow min-cut where minimum cut is a limiting factor, so from above graphic, by splitting the network in disjoint sets, one set is the limiting factor which is the top edge that sets maximum weight of 3, while bottom sets at 9.

Example 5.2

Find the maximum flow for this network?

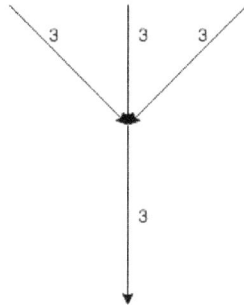

As the limiting factor is on the bottom of network, but weights are same, so maximum flow is still 3. If the network is partitioned by a barrier, it shows that the bottom edge is limiting the flow of the network, so maximum flow will be 3.

5.2. Bipartite Graphs

Bipartite graph is a set of graph vertices which decomposed in two disjoint sets wherre no two graph vertices in same set are adjacent. A bipartite graph is a special case of k-partite graph with k=2. The Figure 5.1 shows bipartite graphs, with vertices in each graph colored as per disjoint sets they belong.

Figure 5.1: Bipartite Graph.

Consider two groups of people signing an agreement. Once both the group signed up, the graph is designed which gives the descriptions about the two groups of people as shown in Figure 5.2. In this graph, the people are selected as per perfect matching features. All of the information is entered into a computer, and computer organizes the information in form of a graph whose vertices are the people having an edge among them that shows if they both be happy to be matched with other person.

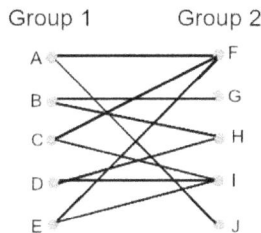

Figure 5.2: Graph of Two Groups of People.

Network Flow Models and Applications

In this graph, there are two groups of vertices (group 1 and group 2) such that vertices which are in similar group have no edges among them. Such type of graphs in mathematics are known as bipartite graph where vertices can be put in two separate groups so that only edges are among those two groups without any edges among the vertices in similar group.

An important concept in graph theory is matching of graph which really useful in various applications of bipartite graphs. In bipartite graph, to represent the member's selections, set of edges are such that there is only one edge for every vertex. Matching of a graph is set of edges in the graph where no two edges shares a common vertex where each vertex has only one edge connected to it in matching (Figure 5.3).

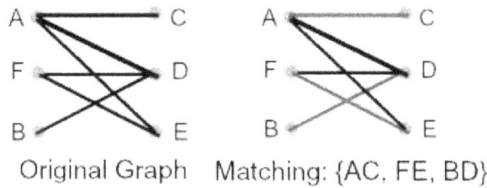

Original Graph   Matching: {AC, FE, BD}

Figure 5.3: Graph Matching.

When a matching is such that on adding edges, there results no matching, then such type of matching is described as maximum matching. Maximum matching is done with maximum number of edges.

5.3 Types of Bipartite Graphs

5.3.1 Acyclic Graphs

Acyclic graph has no graph cycles which are a bipartite graph which can be a tree when connected and is called as forest when disconnected. The numbers of acyclic graphs on n=1, 2, ... are 1, 2, 3, 6, 10, 20, 37, 76, 153, ...and corresponding numbers of connected acyclic graphs are 1, 1, 1, 2, 3, 6, 11, 23, 47, 106, ... (Figure 5.4).

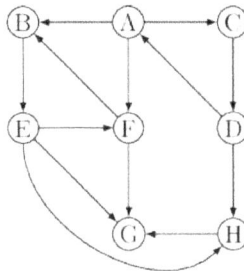

Figure 5.4: Acyclic graph

5.3.2 Book Graphs

The m-book graph is defined as graph Cartesian product $S_{m+1} \square P_2$, where S_m is star graph and P_2 is path graph on two nodes (Figure 5.5).

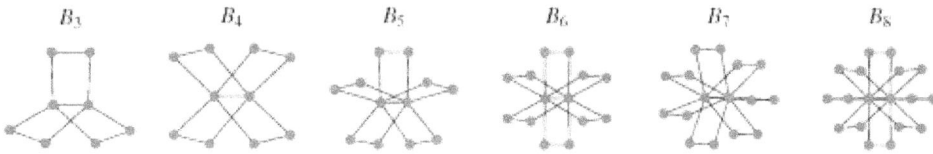

Figure 5.5: m-book Graph.

The general book graph of n stacked pages is (m, n) stacked book graph. Special cases of the m-book graph are:

m	graph
1	square graph
2	domino graph

The n-book graph has chromatic polynomial, independence polynomial, matching polynomial, and rank polynomial given by

$$\pi(x) = (x-1)x\left(x^2 - 3x + 3\right)^n$$

$$I(x) = 2x(1+x)^n + (1+2x)^n$$

$$\mu(x) = (x-1)^{n-2}(x+1)^{n-2}\left[n^2x^2 + \left(x^2 - 1^3\right) + n\left(-1 + 2x^2 - 2x^4\right)\right]$$

$$R(x,y) = \frac{\left[1+3x(x+1)\right]^n (y-x) + x(y+1)\left\{1 + x\left[3 + x(3+y)\right]\right\}^n}{y}$$

5.3.3 Crossed Prism Graphs

An n crossed prism graph for positive even n is a graph generated by taking two disjoint cycle graphs Cn and adding edges (v_k, v_{2k+1}) and (v_{k+1}, v_{2k}) for $k=1,3\ldots, (n-1)$ (Figure 5.6).

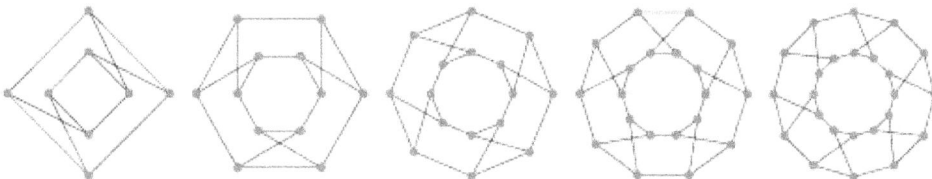

Figure 5.6: n-crossed Prism Graph.

The crossed prism graphs are cubic vertex-transitive, weakly regular, Hamiltonian, and Hamilton-laceable.

5.3.4 Crown Graphs

The *n*-crown graph for an integer n ≥ 3 is the graph with vertex set $\{x_0, x_1, \ldots m\, x_{n-1}, y_0, y_1, \ldots, y_{n-1}\}$ and edge set $\{(x_i, y_j): 0 \le i, j \le n-1, i \ne j\}$ (Figure 5.7).

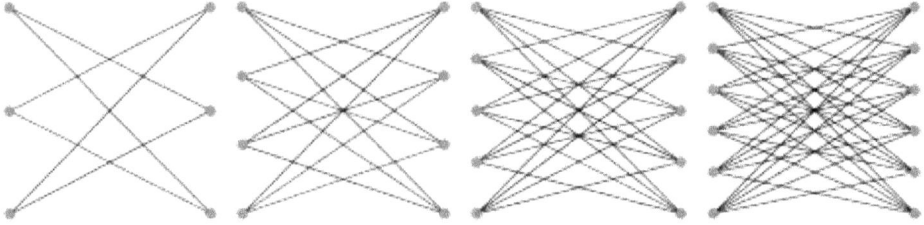

Figure 5.7: Crown Graphs.

It is therefore equivalent to the complete bipartite graph $K_{n,n}$ with horizontal edges removed.

5.3.5 Cycle Graphs of Even Order

Cycle graph C_n is known as n-cycle graph on n nodes having a single cycle through all nodes (Figure 5.8).

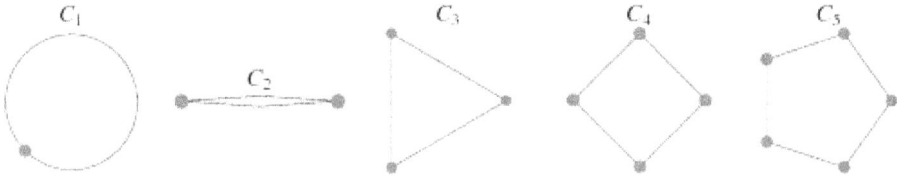

Figure 5.8: Cycle Graphs.

5.3.6 Gear Graphs

Gear graph is a bipartite wheel graph with graph vertex added between each pair of adjacent graph vertices of outer cycle (Figure 5.9).

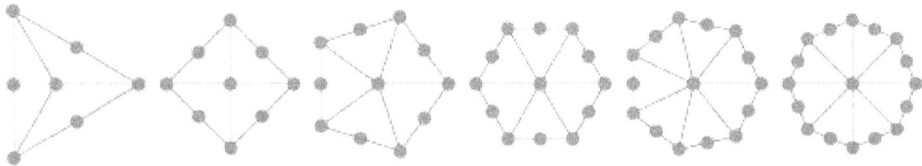

Figure 5.9: Gear Graph.

Gear graph G_n has 2 *n* + 1 nodes and 3 *n* edges. It is proved that all gear graphs are graceful while when two or more vertices are inserted between every pair of vertices of outer cycle of wheel, the resulting graph results appealing.

5.3.7 Grid Graphs

Grid graph is a two-dimensional graph which is also known as square grid graph of order $m \times n$ lattice graph with $G_{m,n}$ where the graph Cartesian product $P_m \square P_n$ of path graphs on m and n vertices. A generalized grid graph can also be defined as $G_{m,n,r}$...(Figure 5.10).

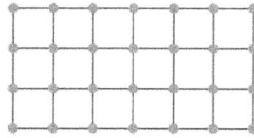

Figure 5.10: Grid Graph.

5.3.8 Haar Graphs

Haar graph $H(n)$ is a bipartite regular vertex-transitive graph which is indexed by positive integer that is generated by binary encoding of cyclically adjacent vertices and these can be connected or disconnected graphs (Figure 5.11).

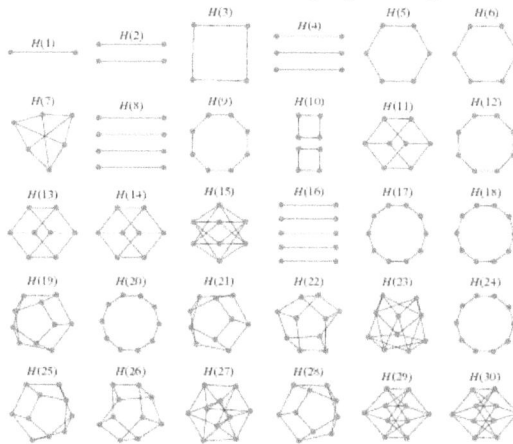

Figure 5.11: Haar Graph.

5.3.9 Hypercube Graphs

An n–hypercube graph is n-cube graph shown by Q_n or 2^n having vertices of 2^K written as $\in_{1,...,} \in_1$ where $\in_1 = 0$ or 1 and two vertices are adjacent if symbols differ in exactly one coordinate (Figure 5.1).

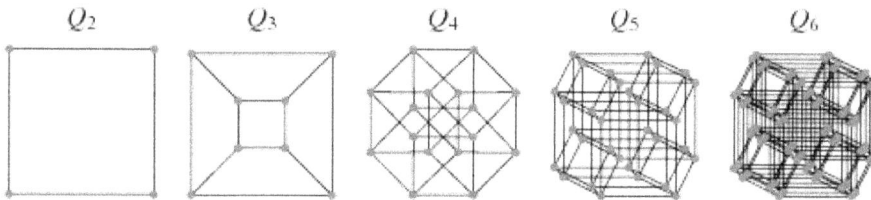

Figure 5.12: Hypercube Graph.

5.3.10 Ladder Graphs

The ladder graph is named for its resemblance to a ladder consisting of two rails and n rungs between them (Figure 5.13).

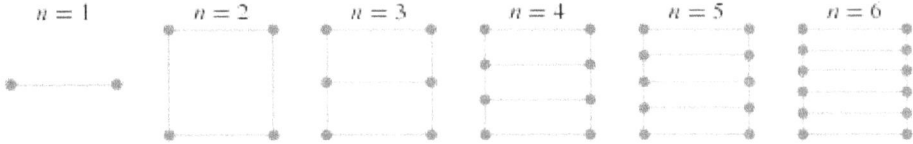

Figure 5.13: Ladder Graph.

5.3.11 Path Graphs

The path graph P_n is a tree with two nodes of vertex degree 1, and the other $n-2$ nodes of vertex degree 2. A path graph is therefore a graph that can be drawn so that all of its vertices and edges lie on a single straight line (Figure 5.14).

Figure 5.14: Path Graph.

5.3.12 Stacked Book Graphs

Stacked book graph are of order (m, n) which is a cartesian product $S_{m+1} \square P_n$, where S_m is star graph and P_n is path graph on n nodes. The graphs corresponds to edges of n copies of an m-page book kept one on top of other which is a generalization of book graph (Figure 5.15).

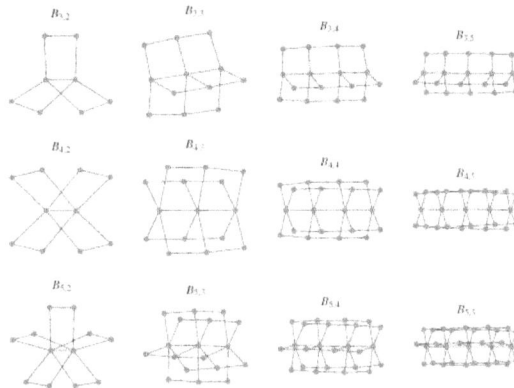

Figure 5.15: Stacjed Book Graphs.

5.4 Matching of Bipartite Graph

Matching designs can be bipartite or non-bipartite. A graph is bipartite if its vertex set can be partitioned into two subsets A and B so that each edge has one endpoint in A and the other endpoint in B (Figure 5.16).

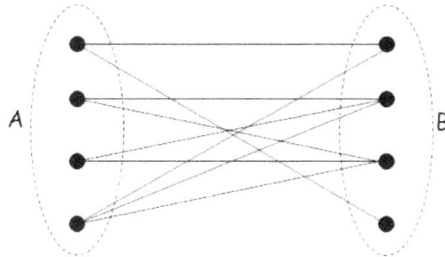

Figure 5.16: Bipartite Graph.

A matching M is a subset of edges so that every vertex has degree at most one in M. In Bipartate matching, replacement of samples without replacement is done while in non-bipartate matching designs involves sampling with replacement. Matching is not the single way, but propensity scores can be used to control confounding.

A matching is a set of edges with no common vertices. Matching with maximum number of edges is a maximum matching. Maximum bipartite matching for a graph isn't unique. No edges in a matching share an end point (Figure 5.17).

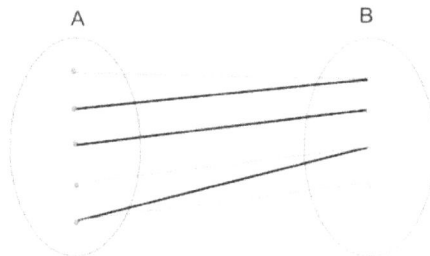

Figure 5.17: Matching.

Matching methods for bipartite matching designs consist of two parts:

➢ Matching ratio
➢ Matching algorithm

The matching ratio can be one-to-one, variable or fixed. The matching algorithm is where the matching actually takes place. One of the most popular algorithms is greedy matching, which includes caliper matching and nearest neighbor matching. Apart from this, other common algorithms are:

➢ **Genetic matching**: iteratively checks the propensity scores and improves them using a combination of propensity score matching and Mahalanobis distance matching.

> ➢ **Optimal matching**: the distance between treated and untreated participants is minimized.

Bipartate matching is also known as conventional two-group matching that involves in creation of pairs from two distinct groups. In bipartate matching, in left side, connections are made between the treatment and controls. This type of matching is equivalent to sampling without replacement (Figure 5.18).

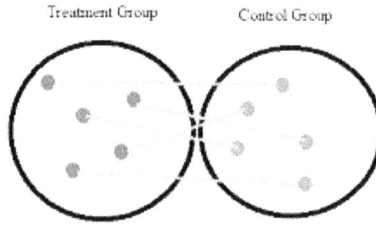

Figure 5.18: Bipartate Matching.

In this, graphically matching of each node is done from treatment group with single node from the control group. A node is a point or circle which is the main unit from which graphs are made. The match is shown by an edge in the graph where an edge serves as a line that connects two nodes. In this, weight is assigned to every edge, which is the difference for some aspect of the pairs such as difference in age, height or BMI.

A good matched pairs normally have very small difference among them, hence smaller weights are preferred. Since the concept is simple, the calculations are not especially if you have multiple covariates. Algorithms like the Greedy Matching Algorithm have been developed to create ideal weights between nodes.

5.4.1 Rules of Matching

The bipartite matching problem is as follows:

Input: two mutually exclusive sets of equal size U and V, along with a list of ordered pairs of the form (u, v) where u U and v V, indicating pairs of members, one of each set that can be "paired" together.

Output: True, if there exists a way to pair up each item in U with an item in V such that each item in both sets appears in exactly one pairing, and false otherwise.

> ➢ Solution: If n = the size of each input set, then:
> ➢ set a graph with 2n+2 vertices
> ➢ create a vertex for each item in each set
> ➢ add source and sink vertices
> ➢ add an edge from source to each item in set U with capacity 1

> ➢ add an edge from each item in set V to sink with capacity 1
> ➢ add an edge between each item in set U and set V that are in set of ordered pairs with capacity 1
> ➢ calculate maximal flow of network

If answer is n, then complete matching exists else complete matching doesn't exist. To match, keep track of each "edge" added during each iteration of the algorithm.

5.4.2 Instance of Bipartite Matching

Set of companies:{Google,Microsoft, Facebook, Amazon, Lockheed}

Set of students: {Amanda, Belinda, Cameron, David, Elyse}

Set of offers: {(Google, Belinda), (Google, Elyse), (Microsoft, Amanda), (Microsoft, Cameron), (Facebook, Amanda), (Facebook, David), (Facebook, Elyse), (Amazon, Amanda), (Amazon, Elyse), (Lockheed, Belinda) }

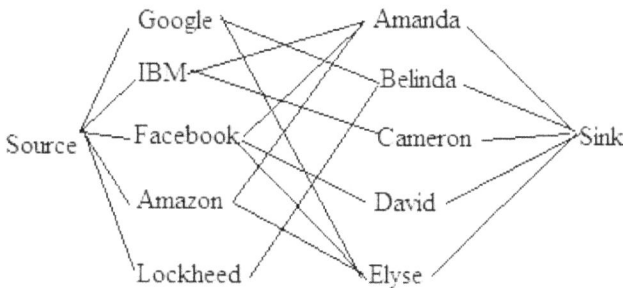

Second Instance of Bipartite Matching

Set of boys: {Bob, Dave, Fred, Harvey}

Set of girls: {Alice, Carol, Elaina, Giselle}

Set of ordered pairs: { (Bob, Elaina), (Bob, Giselle), (Dave, Elaina), (Fred, Alice), (Fred, Carol), (Fred, Giselle), (Harvey, Giselle) }

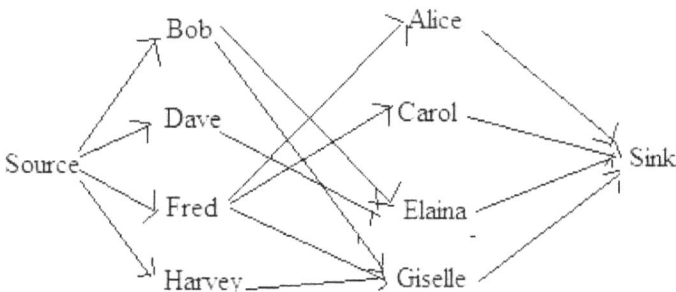

Grand Dinner Problem

N teams attend a dinner. Team i has t_i members. There are M tables at the dinner, with $M \geq N$. Table i can has s_i chairs. We wish to seat all teams such that no two team members are at the same table, so that we maximum students getting to meet members of other teams. Can we do so?

Solution:

Create a flow network with N + M + 2 vertices. Create one vertex for each team and one for each table. Create extra source and sink vertices. Create edges from the source to each team with a capacity of t_i. Create edges from each table vertex to the sink vertex with capacity s_i. Finally, add edges from each team to each table, with capacity 1, since each team can provide at most one person per table. Run the network flow algorithm. If the maximal flow equals the sum of the number of team members, the seating can be done. Otherwise, it can not be.

Two small examples:
team sizes: 4, 5, 3, 5
table sizes: 3, 5, 2, 6, 4
team sizes: 4, 5, 3, 5
table sizes: 3, 5, 2, 6, 3

Museum Guard Problem

A museum employs guards that work in 30 minute shifts: 12am - 12:30am, 12:30am-1am, ..., 11:30pm-12 am. Each guard has a list of times he/she can NOT work. Let a particular guard might not be able to work from 3:30 am to 7:30am and from 4:29pm to 8:01pm, so the guard could work in 3am-3:30am and 7:30am-8am shifts, but not in 4pm - 4:30pm shift or 8pm-8:30pm shift. Further every guard has maximum number of hours to work in a day. Find the maximum number of guards to be scheduled in order to cover each shift without violating any of the constraints.

Solution

As usual, we create an extra source and sink.

Each guard is a vertex in flow network. From source, connect an edge to each guard with integer representing maximum number of shifts in which the guard can work.

Each shift is a node in flow network. Add edge with capacity 1 from each guard to each shift where the guard is able to work that shift.

Now set the capacities from shifts - sink to 1, and run the network flow algorithm to get 48. It is known that we can post 1 guard for each slot. Now again running an algorithm with all capacities set to 2 and see if net flow is 96 or not. If so, move onto 3 and so forth. As there are 50 guards, then the answer will never exceed 50 and this will run in time.

Also, try capacities at 25 for each of edges. If this doesn't work, go to 12 and it works, then go to 37 and so on. Instead of checking if 1 works, then 2, then 3, etc, a binary search will hone in on the answer a bit more quickly, especially in the cases that the answer is closer to 50.

Cow Steeplechase Problem

Given a list of horizontal line segments (none of which intersect each other) and vertical line segments (none of which intersect each other), calculate the minimum number of line segments that must be removed, so that no two lines intersect each other, or alteratively, the most number of line segments that mutually don't intersect one another.

Solution

We can create a bipartite matching solution. Our goal is to match horizontal line segments to vertical line segments in such a way that each pair intersects. We know that if we have a set of these intersecting pairs, at the very least, one item in each pair must be removed. Thus, what we really want is the maximum matching. Once we have this, then we have proof that 7 of the segments must be removed. No other matching forces us to remove more. Thus, that how many we are forced to remove to create no intersecting line segments. Alternatively, we can calculate the maximum set of segments we can have without any two intersecting by taking the total number and subtracting out this maximum matching.

5.5 Non-Bipartite Matching

Non-bipartate design is also known as multi-group matched design which produces pairs from multiple groups. It is similar to sampling with replacement. The Bipartite designs are more common, but non-bipartite designs are available for exceptional case where a member needs to be reused. In such case, similar matching control is applied for two or more treatment group participants (Figure 5.19).

Figure 5.19: Non-Bipartate Design.

In above Figure, every node in the right hand box is in a separate group. Non-bipartite designs are available for when you want to reuse a member. For example, if same control is applied as matching for two or more treatment group, then an augmenting path algorithms such as Blossom V algorithm are used for creating non-bipartite matches. They are technically complex, which may be a reason why biparite matches are often preferred.

In case of non-bipartite designs, matching procedure are hard and the usual algorithm is bootstrapping which involves drawing bootstrap samples. The matching algorithm step is usually performed with software.

5.5.1 Propensity Score

Propensity score is probability that a unit with certain characteristics will be assigned to the treatment group. The scores can be used to reduce or eliminate selection bias in observational studies by balancing covariates among treated and control groups. In case of covariates are balanced, it become much easier to match participants with multiple characteristics (Figure 5.20).

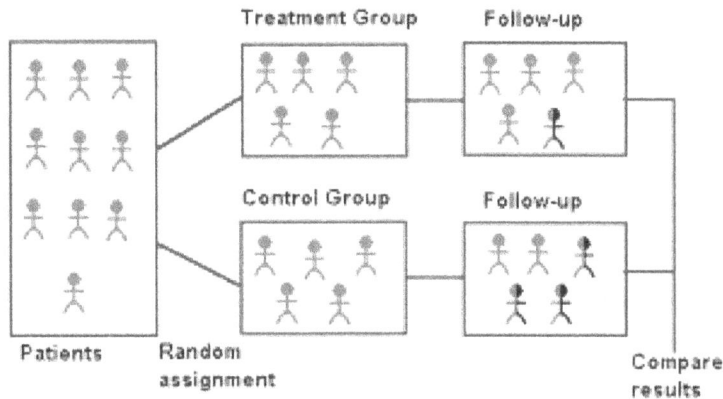

Figure 5.20: Propensity Score Matching.

Propensity score matching generates sets of participants for treatment and control groups. A matched set comprises of single participant in treatment group and one in control group with similar propensity scores. The idea of this is to approximate a random experiment, eliminating many of the problems that come with observational data analysis.

Steps involved in propensity score matching are:

- ➤ Collection and preparing data.
- ➤ Estimating propensity scores where correct scores are unknown which are estimated by methods such as discriminant analysis, logistic regression, and random forests.
- ➤ Matching of participants using estimated scores.
- ➤ Evaluating covariates for even spread across groups where scores are good estimates for true propensity scores when matching process successfully distributes covariates over the treated/untreated groups.

5.6. Maximum Bipartite Matching

Matching in Bipartite Graph is a set of the edges selected in such a way that no two edges share an endpoint. A maximum matching is a matching of maximum size. In maximum matching, if any edge is added, it is no longer a matching. There can be more than one maximum matchings for a given Bipartite Graph.

Matching of maximum cardinality is called as maximum matching where a graph may have more than one matching of maximum cardinality. In the graph shown below, it is difficult to get a matching with more than 3 edges as there are only 6 vertices where for each pair of vertices there can be at most one edge in matching, so there cannot be a matching with greater than 3 edges (Figures 5.21 and 5.22).

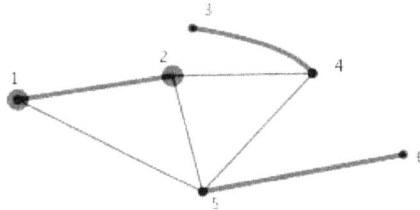

Figure 5.21: { (1,2), (3,4), (5,6) } Form a Maximum Matching.

Hence, this is a maximum matching.

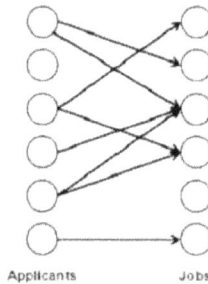

Figure 5.22: Graph of Applicants vs Job.

There are many real world problems that can be formed as Bipartite Matching. If there are M job applicants and N jobs where every applicant has subset of jobs in which he/she is interested in (Figure 5.23).

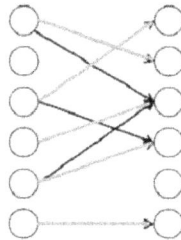

Figure 5.23: Probability of Applicants Getting Jobs.

Each job opening can only accept one applicant and a job applicant can be appointed for only one job, then an assignment of jobs to applicants is such that as many applicants will possibly get the jobs. This is Maximum Bipartite Matching problem which can be solved by converting to flow network by following the particular steps (Figure 5.24):

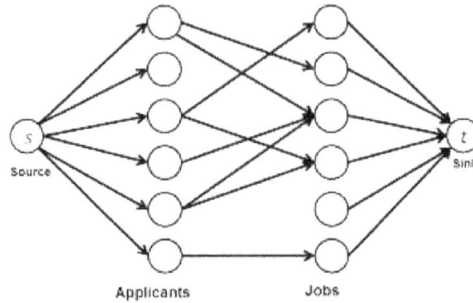

Figure 5.24: Building a Flow Network.

Building a Flow Network: There must be a source and sink in a flow network. So we add a source and add edges from source to all applicants. Similarly, add edges from all jobs to sink. The capacity of every edge is marked as 1 unit.

Finding maximum flow: With the use of Ford-Fulkerson algorithm maximum flow in flow network can be created. The maximum flow is actually the MBP which can be solved by first defining the input and output forms where input is in Edmonds matrix that is a 2D array 'bpGraph[M][N]' having M rows and N columns hose value is 1 if i'th applicant is interested in j'th job, otherwise 0.

Here output is number maximum number of people that can get jobs. For this, create a matrix that shows adjacency matrix of direct graph with M+N+2 vertices and call fordFulkerson() for the matrix whose implementation needs $O((M+N)*(M+N))$ extra space.

Extra space can be reduced and code can be simplified as graph is bipartite whose capacity of every edge is 0 or 1. Using DFS traversal, job for an applicant can be located by calling bpm() for each applicant where bpm() is DFS based function with all possibilities that assigns job to the applicant.

In case of bpm(), by trying all jobs which applicant 'u' is interested till a job is located or by luck. If a job is not assigned to any applicant, then assign it to applicant and return true. If a job is assigned to applicant else say x and recursively check whether x can be assigned some other job. For making this correct, x doesn't get the same job again, so mark job 'v' as seen before by recursive calling for x. If x gets other job, change applicant for job 'v' and return true. With an array maxR[0..N-1], stores the applicants assigned to different jobs. If bmp() returns true, then it shows an augmenting path in flow network and 1 unit of flow is added to result in maxBPM().

Example

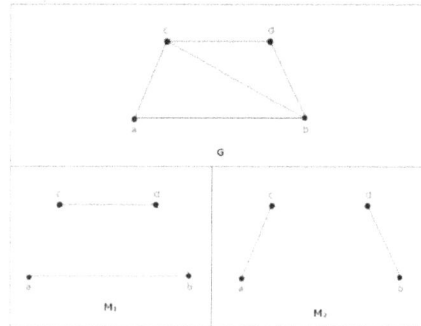

For a graph given, M1 and M2 are the maximum matching of 'G' and its matching number is 2. Hence by using the graph G, we can form only the subgraphs with only 2 edges maximum. Hence we have the matching number as two.

Example

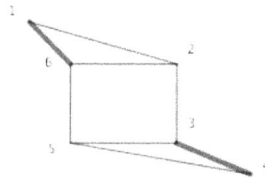

Maximal matching is that where no edge can be added to matching { (1,6), (3,4) } without violating the matching property. The size of particular maximal matching of G is 2.

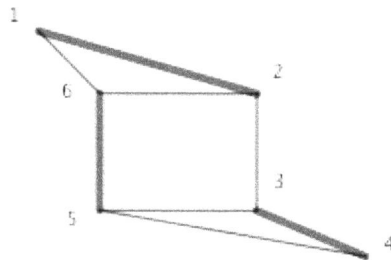

A Maximum matching for G: { (1,2), (3,4), (5,6) }where above matching perfect matching of G as all vertices are paired.

Example

Find a matching with the maximum number of edges.

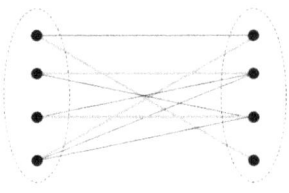

It is seen that perfect matching is that matching in which every vertex is matched.

Step 1: Greedy method

Add an edge with both endpoints unmatched

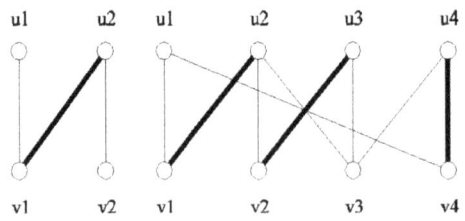

Step 2: Hall's Theorem

As per Hall's Theorem, a bipartite graph G=(A,B;E) has matching that saturates A if and only if |N(S)| >= |S| for every subset S of A.

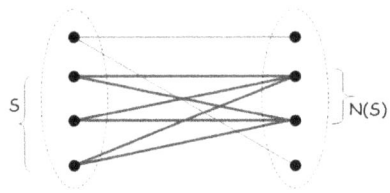

In a bipartite graph, the size of a maximum matching is equal to the size of a minimum vertex cover.

Step 3: Augmenting Path

Given a matching M, an M-alternating path is a path that alternates between edges in M and edges not in M. An M-alternating path whose endpoints are unmatched by M is an M-augmenting path.

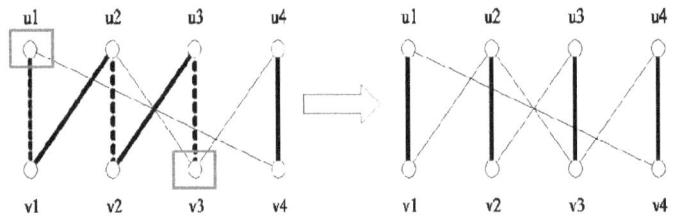

$M^* = m \oplus P$

M-augmenting paths

Orient the edges in M go up while others go down. An M-augmenting path is a directed path between two unmatched vertices

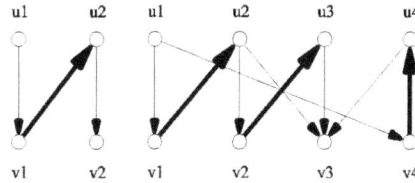

5.6.1 Perfect Matching

A perfect matching of a graph is a matching (*i.e.*, an independent edge set) in which every vertex of the graph is incident to exactly one edge of the matching. A perfect matching is therefore a matching containing n/2 edges (the largest possible), meaning perfect matchings are only possible on graphs with an even number of vertices (Figure 5.25).

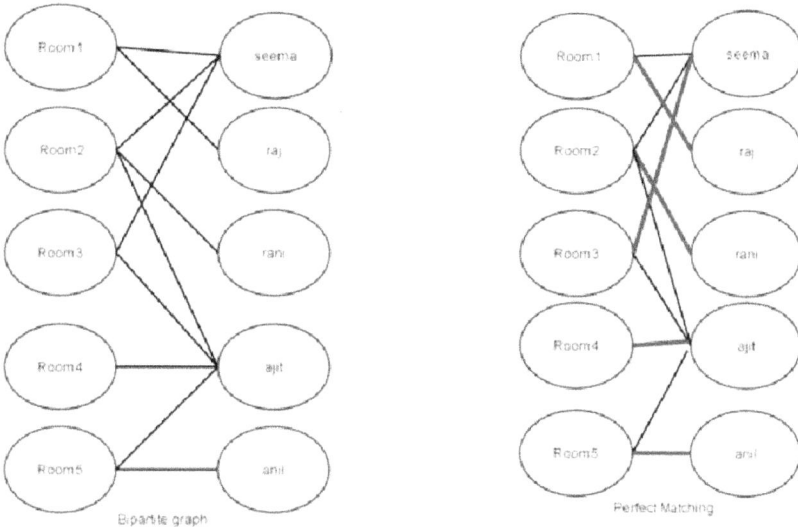

Figure 5.25: Bipartite and Perfectly Matched Graphs.

Matching in a graph G=(V,E) is subset M of edges E such that no two edges in M share a common end node. A perfect matching M in G is matching such that each node of G is incident to an edge in M. If edges of graph have associated weight, then maximum/minimum weighted perfect matching is perfect matching where sum of weights of edges in matching is maximum/minimum.

Algorithms for finding Perfect Match

There are three algorithms for doing perfect matching of bipartite graph:

Max Flow:

After determining the two halves, add super-source to one the sets and a super-sink to the other, ensuring all edges have capacity one. From the definition of Max-Flow, push as much flow as possible to maximize in maximal matching. This algorithm is dependent on how fast the Max-Flow algorithm is, and it tends to be slower. Moreover, to retrieve the set of matching's, a Breadth-First Search Traversal of the resulting graph needs to be done.

Augmenting path algorithm:

The idea is to find an augmenting path and flip the parity noting that it increase the number of edges matched. This gives a O(VE) algorithm.

Hopcroft-Karp:

This is efficient version of augmenting path algorithm where instead of finding any augmenting path, it finds the most efficient augmenting path at any time. Such properties causes algorithm to speedup to $O((\sqrt{V})E)$ time.

Example

In the following graphs, M_1 and M_2 are examples of perfect matching of G.

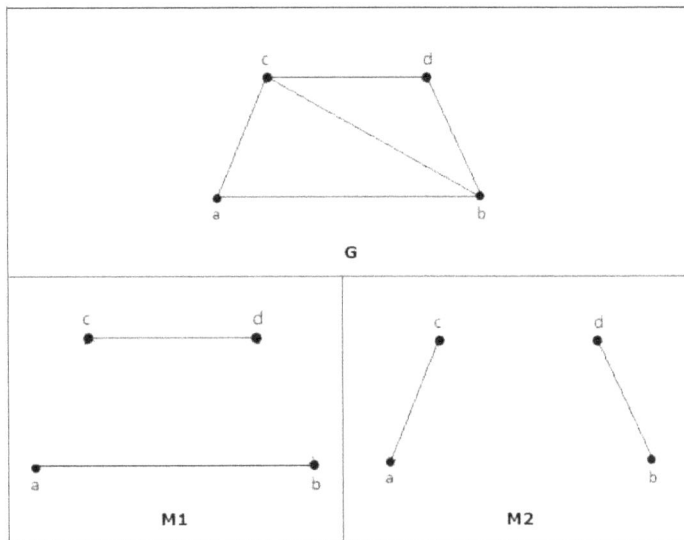

It is observed that every perfect matching of graph is a maximum matching of graph as there is no chance of adding one more edge in a perfect matching graph. A maximum matching of graph need not be perfect. If graph 'G' has perfect match, then number of vertices |V(G)| is even while if it is odd, then last vertex pairs with other vertex and hence there remains a single vertex that cannot be paired with any vertex for which degree is zero. It means that it is clearly violating the perfect matching principle.

Example

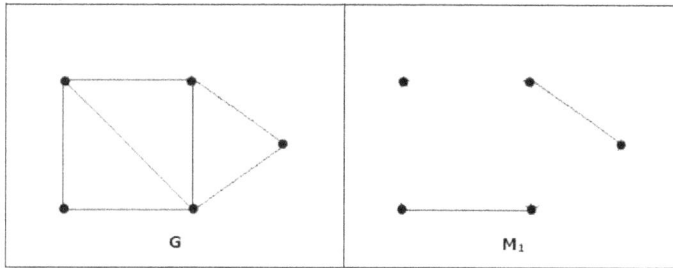

It is observed that converse of statement shown above need not be true when G has even number of vertices, for M_1 need not be perfect.

Example

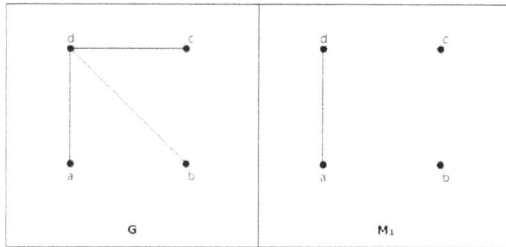

It is matching, but it is not a perfect match, even though it has even number of vertices.

5.7 k-Regular Bipartite Graphs

Regular graph is a graph where each vertex has similar number of neighbors with vertex having same degree or valency. A regular directed graph should also satisfy stronger condition that the in-degree and out-degree of each vertex results equal to each other. A regular graph with vertices of degree k is k-regular graph or regular graph of degree k. The regular graph of odd degree will contain an even number of vertices.

Regular graphs of degree with 2 are easy to classify where 0-regular graph has disconnected vertices, 1-regular graph has disconnected edges while 2-regular graph has disconnected cycles and infinite chains.

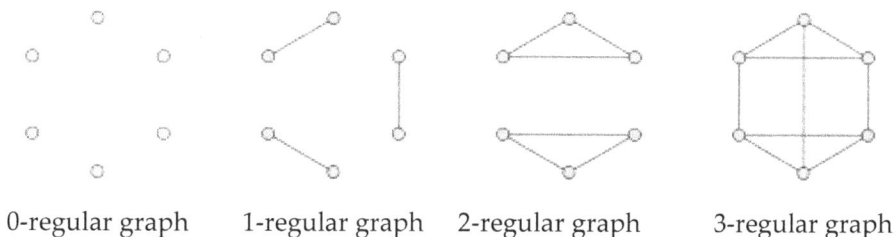

0-regular graph   1-regular graph   2-regular graph   3-regular graph

5.8 Disjoint Paths

Two or more paths are arc-disjoint when they have no common arcs. It is noted that node-disjoint paths are arc-disjoint paths. It is considered that disjoint paths problem is an extension of the shortest path problem where instead of having single shortest path, many paths that do not share any common links can be find out.

In case of disjoint paths to network traffic results an increase in reliability of network connections with network survivability. Network survivability is network's capability which shows continuous service in presence of network component failures. Other variant of disjoint paths problem is disjoint path pairs problem in which, rather than finding many disjoint paths for a single pair of source and destination nodes, a single path is calculated for every pair of source and destination nodes, such that these paths are disjoint.

Disjoint paths have many applications such as with multiple disjoint paths for traffic in a communication network, there are chances of improvement of transmission reliability. By sending traffic concurrently on multiple disjoint paths, failure of path will have no affect on performance of paths, and traffic still reach its destination. In transportation networks, having a pre-calculated number of disjoint paths would enable a truck driver to follow a different path for a change of scenery instead of always sticking to the shortest path.

There are certain conditions that could be imposed on disjoint paths like:

> ➢ Min-max disjoint paths problem where sum of weights of constituent links of path with largest path weight is minimized.

> ➢ Min-min disjoint paths problem where sum of weights of constituent links of path with smallest path weight results minimized.

> ➢ Bounded disjoint paths problem where sum of weights of constituent links of each path be less than D.

> ➢ Min-sum disjoint paths problem here sum of weights of constituent links of k paths is minimized.

> ➢ Consider K Shortest Arc-Disjoint Paths Problem where it is required to find a set $\{p_1,...,p_k\}$ of arc-disjoint paths which is equivalent to minimal cost flow problem as shown:

$$\text{minimize } \Sigma_A c_{ij} \times X_{ij}$$

$$\Sigma_j X_{ij} - \Sigma_j X_{ji} = \begin{cases} K \ if \ i = s \\ 0 \ if \ i \neq s, t (\text{for any } i \in N), \\ -K \ if \ i = t \end{cases}$$

$$0 \leq x_{ij} \leq 1, \ \left(\text{for any } (i,j) \in A\right).$$

Here it is required to find a feasible flow value K from initial node to terminal node such that total cost will be minimal. The flow in each arc cannot exceed the unity. In case, when such a flow does not exists, then similar can be concluded about K disjoint paths where optimal solution of minimal cost flow problem will not show about k-th shortest disjoint path. If it is required to know about k-th shortest disjoint paths, then it needs framing of new network whose set of arcs is set of arcs for flow one.

To have shortest disjoint path, second disjoint shortest, third disjoint shortest, and so on to K-th shortest disjoint path, where execution of K times the steps shown: Finding shortest path in resulting network

Removing Arcs

It is seen that an algorithm is exemplified with network on left, where initial node (s) is 1 and terminal node (t) is 4 with the number close to each arc showing cost or distance.

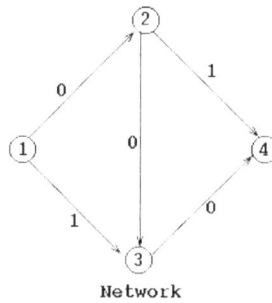

Network

In order to find two disjoint shortest paths from s to t, then as per algorithm, it is required to determine the minimal cost flow with value 2 from s to t, in a network. For optimal solution of the problem, flow is 1 in every arc but (2,4) where is zero, hence resulting network is shown below where shortest path from 1 to 4 is {1, (1,2), 2, (2,4), 4}. After the path is removed, a second shortest disjoint path {1, (1,3), 3, (3,4), 4} is obtained.

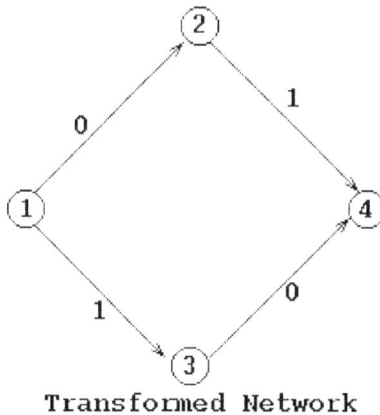

Transformed Network

Here, shortest path in network is {1, (1,2), 2, (2,3), 3, (3,4), 4} which is not one of the two shortest disjoint paths, so it gives incorrect conclusion on solving the problem and executing K times with computation of shortest path and removal of arcs from network.

On considering K Shortest Node-Disjoint Paths Problem where it is required to find a set {p1, ...pk} of node-disjoint paths. Such problem can be transformed in K Shortest Arc-Disjoint Paths Problem by considering node x that belongs to single path. Now splitting node x in x' and x'' nodes that are linked by zero cost arc (x',x'') with arc distances by changing arcs (i,x) to arcs (i,x') and all arcs (x,j) to arcs (x'',j). It is only required to find the K Shortest Arc-Disjoint Paths in resulting network so as to find the K Shortest Node-Disjoint Paths in network shown above.

Example

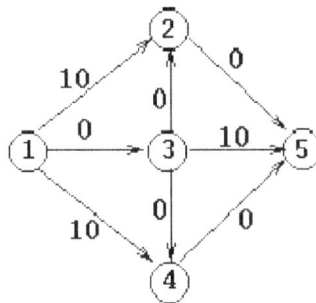

Network

Consider the network shown where it is required to determine three shortest paths from s=1 to t=5, in such a way that no more than one path passes throughout node 3 as shown in figure below as transformed network.

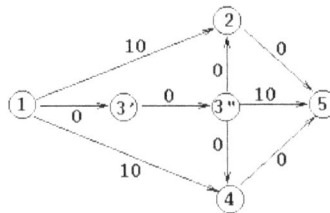

Transformed Network

If we bound to one the flow capacity of all the arcs and we send three units of flow from node 1 to node 5 minimizing the cost, we will get one unit of flow in all the arcs but (3'',4) and (3'',2), where the flow is zero. That is, an optimal solution for the problem would be the paths {1, (1,2), 2, (2,5), 5}, {1, (1,3), 3, (3,5), 5} and {1, (1,4), 4, (4,5), 5}, all of them with cost 10.

5.8.1 Edge Disjoint Paths

An easiest applications of maximum flow is finding maximum number of edge disjoint paths between two specific vertices s and t in directed graph G with maximum flow. A set of paths in G is edge disjoint when every edge in G appears in one of paths where many edge disjoint paths will pass through similar same vertex.

On giving each edge capacity 1, maxflow from s to t will assign a flow of 0 or 1 to every edge. As any vertex of G lies on many two saturated edges, subgraph S of saturated edges is union of many edge disjoint paths and cycles. Also, number of paths will be equal to value of flow, so extracting actual paths from S will be simple as it needs following of directed path in S from s to t, removing that path from S, and recurse.

In a directed graph with two vertices, source 's' and destination 't', the maximum number of edge disjoint paths from s to t will be there when they don't share any edge.

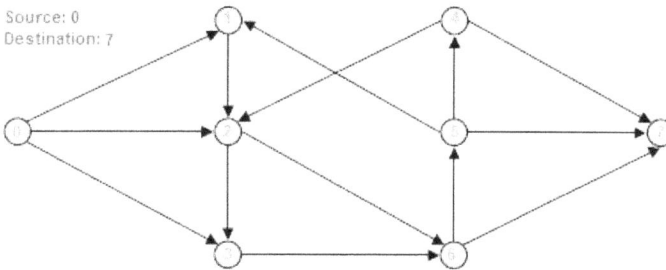

Source: 0
Destination: 7

It is noted that there can be maximum two edge disjoint paths from source 0 to destination 7 in the graph shown above. Also in the graph shown below, two edge disjoint paths are shown as 0-2-6-7 and 0-3-6-5-7.

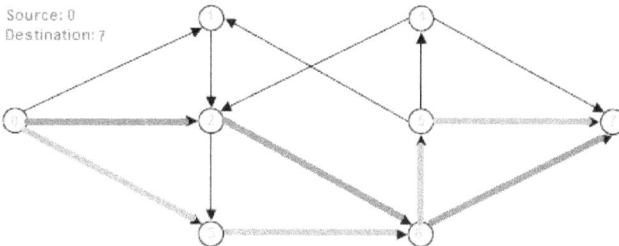

Source: 0
Destination: 7

It is observed that the paths here will be different with maximum number. Also, in above figure, possible set of paths will be 0-1-2-6-7 and 0-3-6-5-7. In order to solve such problem, reduce it to maximum flow problem by considering following steps:

1. Consider the given source and destination as source and sink in flow network. Assign unit capacity to each edge.

2. Run Ford-Fulkerson algorithm to find the maximum flow from source to sink.

3. The maximum flow is equal to the maximum number of edge-disjoint paths.

On running Ford-Fulkerson algorithm, it is observed that the capacity is reduced by unit, hence an edge cannot be used again, so maximum flow will be equal to maximum number of edge-disjoint paths.

5.8.2 Baseball Elimination

In baseball elimination problem there are n teams competing in a league. At a specific stage of the league season, w_i is the number of wins and r_i is the number of games left to play for team i and r_{ij} is the number of games left against team j. A team is eliminated if it has no chance to finish the season in the first place. The task of the baseball elimination problem is to determine which teams are eliminated at each point during the season. As proposed, it is a method which reduces this problem to maximum network flow where a network is created to find whether team k is eliminated or not.

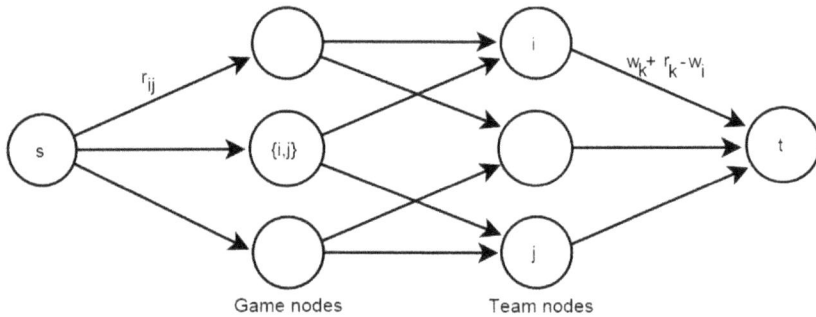

Game nodes   Team nodes

If $G = (V, E)$ a network with $s,t \in V$ as source and sink, then on adding a game node $\{i,j\}$ with $i < j$ to V, and connecting each of them from s by an edge having capacity r_{ij} shows the number of plays among the two teams. Also by adding a team node for every team and connecting each game node $\{i,j\}$ with two team nodes i and j to make sure that one of them wins. Hence the edges are made from team node i to sink node t and capacity of $w_k+r_k-w_i$ is set to save team i from winning more than w_k+r_k. If S is the set of all teams that are taking part in the league with $r(S - \{k\})r(S-\{k\})=\sum i,j\in\{S-\{k\}\},i<jrij\{\displaystyle \scriptstyle r(S-\{k\})=\sum_{\{i,j\in\{S-\{k\}\},i<j\}}r_{\{ij\}} = \sum ij \varepsilon (S - \{k\}) i>j$, then by this method, team k is not eliminated till a flow value of size $r(S - \{k\})$ exists in network G which clearly describes the flow value is the maximum flow value from s to t.

6 | Problem Solving with Network Flow

6.1 Grand Dinner Problem

N teams attend a dinner. Team i has ti members. There are M tables at the dinner, with M ≥ N. Table i can has si chairs. We wish to seat all teams such that no two team members are at the same table, so that we maximum students getting to meet members of other teams. Can we do so?

Solution using Network Flow

Create a flow network with N + M + 2 vertices. Create one vertex for each team and one for each table. Create extra source and sink vertices. Create edges from the source to each team with a capacity of ti. Create edges from each table vertex to the sink vertex with capacity si. Finally, add edges from each team to each table, with capacity 1, since each team can provide at most one person per table. Run the network flow algorithm. If the maximal flow equals the sum of the number of team members, the seating can be done. Otherwise, it can not be.

Two small examples:

team sizes: 4, 5, 3, 5
table sizes: 3, 5, 2, 6, 4

team sizes: 4, 5, 3, 5
table sizes: 3, 5, 2, 6, 3

Example

Each team participating in this year's ACM World Finals is expected to attend the grand banquet arranged for after the award ceremony. To maximize the amount of interaction between members of different teams, no two members of the same team will be allowed to sit at the same table. Given the number of members on each team (including contestants, coaches, reserves, and guests) and the seating capacity of each table, determine whether it is possible for the teams to sit as described. If such an arrangement is possible, output one such seating assignment. If there are multiple possible arrangements, any one is acceptable.

The input file may contain multiple test cases. The first line of each test case contains two integers, 1<=M<=70 and 1<=N<=50, denoting the number of teams and tables, respectively. The second line of each test case contains M integers, where the ith integer m_i indicates the number of members of team i. There are at most 100 members of any team. The third line contains N integers, where the jth integer n_j, 2<=n_j<=100, indicates the seating capacity of table j. A test case containing two zeros for M and N terminates the input.

For each test case, print a line containing either 1 or 0, denoting whether there exists a valid seating arrangement of the team members. In case of a successful arrangement, print M additional lines where the ith line contains a table number (from 1 to N) for each of the members of team i.

Sample Input

```
4 5
4 5 3 5
3 5 2 6 4
4 5
4 5 3 5
3 5 2 6 3
0 0
```

Sample Output

```
1
1 2 4 5
1 2 3 4 5
2 4 5
1 2 3 4 5
0
```

6.2 Max-Flow = Min-Cut

A flow network is a directed graph G = (V, E) where each edge (u, v) has a capacity c(u, v) ≥ 0, and:

➤ If (u, v) ∉ E then c(u, v) = 0.

➤ If (u, v) ∈ E then reverse edge (v, u) ∉ E. [*]

➤ A vertex s is designated as the source vertex.

➤ A vertex t is designated as the sink vertex (or t for "target").

Example: Trucking capacity network:

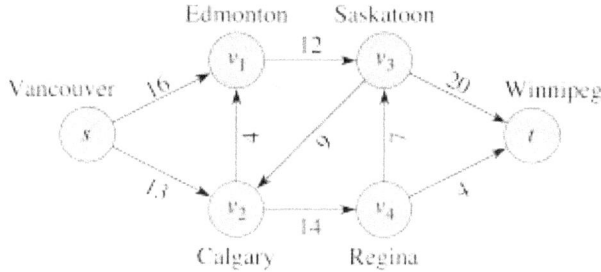

A flow for a network is a function $f : V \times V \to \Re$ where f assigns numbers to edges that satisfies:

Capacity Constraint: $\forall\ u, v \in V, 0 \leq f(u, v) \leq c(u, v)$.

Flow Conservation: $\forall\ u \in V - \{s, t\}$,

$$\sum_{v \in V} f(v, u) = \sum_{v \in V} f(u, v)$$

Example: Flow/capacity:

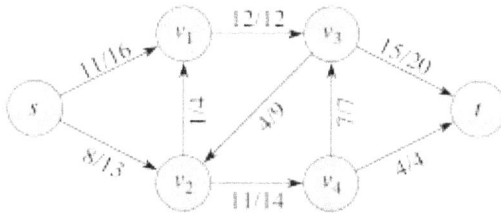

The value of flow $f = |f|$ is flow out of source which is minus the flow in source:

$$|f| = \sum f(s, v) - \sum f(v, s)$$

It is noted that the maximum flow min-cut theorem is network flow theorem which shows that the maximum flow through any network from a given source to given sink is sum of edge weights which if removed will disconnect source from sink. Also, for any network graph and a selected source and sink node, max-flow from source to sink = min-cut required to separate source from sink.

Max-flow min-cut has many applications. In computer science, networks depends on algorithm, so network reliability, availability and connectivity is more. In mathematics, matching in graphs uses such algorithm.

Example

The capacity of cut (ii) for a network as shown in figure has already found as 22. The figures circled will show a way of getting the flow and as per the theorem, it should be maximum possible flow through the network.

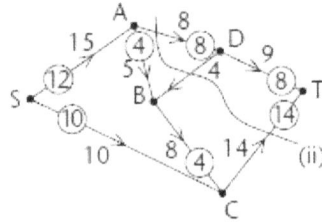

(ii)

Cuts and Flow

A cut (S, T) of a flow network G = (V, E) is a partition of V into S and T = V - S such that s ∈ S and t ∈ T.

In an example, the net flow across cut (S, T) for flow f is:

$$f(S,T) = \sum_{u \in S} \sum_{v \in T} f(u,v) - \sum_{u \in S} \sum_{v \in T} f(v,u)$$

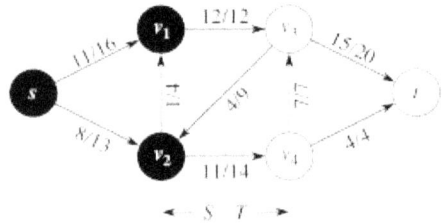

The capacity of cut (S, T) is:

$$c(S,T) = \sum_{u \in S} \sum_{v \in T} c(u,v)$$

Here the asymmetry among net flow and capacity of cut, edges are only counted as going from S to T and ignoring in reverse direction. For total flow, count flow on edges across the cut will be the flow on edges from S to T which is minus the flow on edges from T to S.

Examples

Consider the cut S = {s, w, y}, T = {x, z, t} in network shown.

$$f(S.T) = \underbrace{f(w.d) + f(y.z)}_{\text{from S to T}} - \underbrace{f(x.y)}_{\text{from T to S}}$$

$= 2 + 2 - 1$

$= 3.$

$$c(S.T) = \underbrace{c(w.x) + c(y.z)}_{\text{from S to T}}$$

$= 2 + 3$

$= 5.$

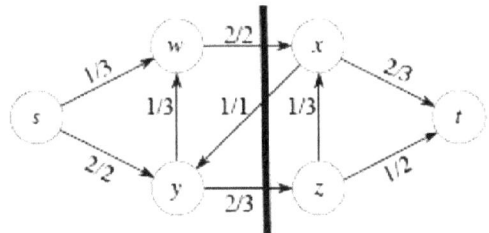

Now consider the cut S = {s, w, x, y}, T = {z, t}.

$$f(S.T) = \underbrace{f(x.d) + f(y.z)}_{\text{from S to T}} - \underbrace{f(z.x)}_{\text{from T to S}}$$

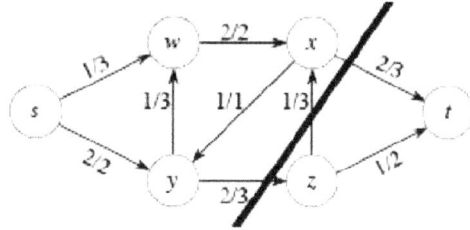

$$= 2 + 2 - 1$$

$$= 3.$$

$$c(S.T) = \underbrace{c(x.t) + c(y.z)}_{\text{from S to T}}$$

$$= 3 + 3$$

$$= 6.$$

Minimum Cut

A minimum cut of G is a cut whose capacity is minimum over all cuts of G. The proofs of these are straightforward but involve long manipulations of summations: see text.

For any cut (S, T), f(S, T) = |f|
(the net flow across any cut equals the value of the flow).

The intuition is that no matter where you cut the pipes in a network, you'll see the same flow volume coming out of the openings. If you did not, conservation would be violated at some nonempty subset of the vertices.

Here the value of any flow ≤ capacity of any cut. It is intuitive under plumbing analogy as if it is incorrect, then push more flow through the pipes that can be hold.

6.3 Image Segmentation Problems

Image segmentation is a task of distinguishing objects from background in unseen images. Typically this division is based on low level cues such as intensity, homogeneity or contours. It works with certain popular approaches such as threshold techniques, edge-based methods, region-based techniques and connectivity- preserving relaxation methods.

Graph Cuts

A graph cut is process of partitioning a directed or undirected graph in disjoint sets. The concept of optimality of such cuts is introduced by associating an energy to each cut. Nevertheless, ever since it became apparent that many low-level vision problems can be posed as pending energy minimizing cuts in graphs these techniques have received a lot of attention in the computer vision community. Graph cut methods have been successfully applied to stereo, image restoration, texture synthesis and image segmentation.

The method is used in computer vision to locate objects boundaries. It can be applied precisely to cut out objects from an image database or can be applied as real-time like in robotics application. The idea of object segmentation is to locate boundary among an object and background. It was proved that solving the minimal cut problem is equivalent to finding the maximum amount which can flow from source to sink.

It will saturate the network at same edge which is cut by the minimal cut. There are many algorithms from combinatorial optimization domain which solve the maximum flow problem and thus our problem. The variant presented is tuned to solve the problem on practical graphs from computer vision. It uses the fact that our graph is regular, each node is a pixel with only four neighbors.

Algorithm:

The edge capacities are initialized from the difference between pixels. The algorithm then progresses in three step:

> ➢ Grow step: Active nodes explore their neighboring and parent free nodes until they encounter a node from the other tree. When two nodes from opposite trees meet, they form a path.

> ➢ Augment step: We find the bottleneck along the found path and push the maximum flow through the path.

> ➢ Adoption step: Some edge will have been saturated, thus cutting parents from their children which tries to find certain parents in their neighbourhood. Here the parents can adopt orphans from similar tree and if connected to their roots.

6.4 Maximal Anti-Chain Problems

Anti-chain is a subset of partially ordered set where any two elements in subset are incomparable. If S is a partially ordered set and a and b are two elements of partially ordered set which are comparable when a ≤ b or b ≤ a and if two elements are not comparable, then they are said to be incomparable where neither x ≤ y nor y ≤ x.

It is seen that chain in S is a subset C of S where every pair of elements is comparable and C is totally ordered. An anti-chain in S is a subset A of S where every pair of different elements is incomparable that has no order relation among any two different elements in A.

Maximal anti-chain is an anti-chain which is not a proper subset of any other anti-chain. A maximum anti-chain is anti-chain having cardinality at least as large as every other anti-chain. Here the width of a partially ordered set is cardinality of maximum anti-chain where any anti-chain can intersect any chain in one element. Hence on partitioning the elements of order in k chains, width of order should be at most k.

As per Dilworth's theorem, bound can always be reached where there exists an anti-chain and partition of elements in chains such that number of chains will be equal to number of elements in the anti-chain that are equal to the width.

Similarly, it is noted that height of a partial order to be maximum cardinality of a chain can be showed by Mirsky's theorem which explains that in any partial order of finite height, height equals the smallest number of anti-chains in which the order may be partitioned.

6.5 Vehicle Routing

Vehicle Routing involves determining the optimal routes that vehicles can take to reach their destinations *e.g.* from factories to retail depots or warehouses. This involves the usage of algorithms to determine efficient routes taking into account order volumes, constraints such as time windows, vehicle requirements, *etc.* It is useful for daily route scheduling and optimisation, strategic transport and planning, and real time fleet management for a variety of businesses and operations in various sectors.

Consider the situation shown below where we have a depot surrounded by a number of customers who are to be supplied from the depot.

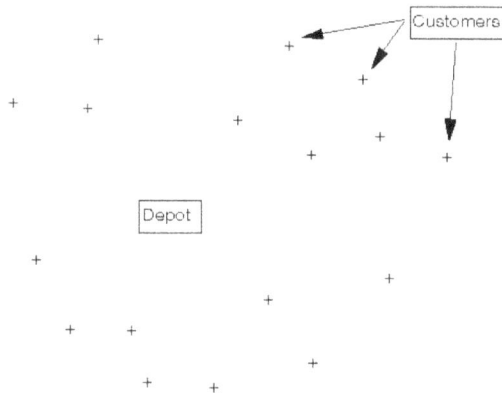

The depot manager faces the task of designing routes (such as those shown below) for his delivery vehicles and this problem of route design is known as the vehicle routing or vehicle scheduling problem.

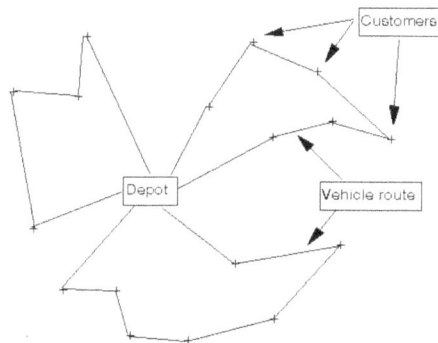

Hence the vehicle routing problem can be defined as the problem of designing routes for delivery vehicles (of known capacities) which are to operate from a single depot to supply a set of customers with known locations and known demands for a certain commodity. Routes for the vehicles are designed to minimise some objective such as the total distance travelled.

This problem is one that has a long history of systematic study, the problem first being considered in an academic paper by Dantzig and Ramser published in the late 1950's.

The problem has attracted a lot of attention in the academic literature for two basic reasons:

> ➢ the problem appears in a large number of practical situations
> ➢ the problem is theoretically interesting and not at all easy to solve.

6.5.1 Vehicle Routing Problem

Vehicle Routing Problem (VRP) is a generic name given to a whole class of problems in which a set of routes for a fleet of vehicles based at one or several depots must be determined for a number of geographically dispersed cities or customers. The objective of the VRP is to deliver a set of customers with known demands on minimum-cost vehicle routes originating and terminating at a depot (Figure 6.1).

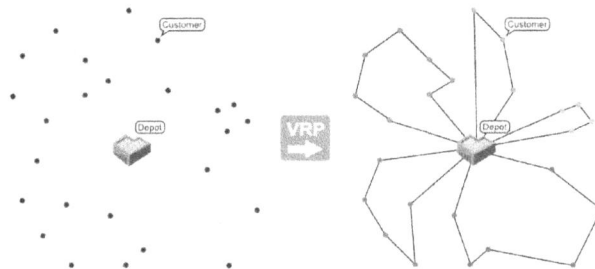

Figure 6.1: Instance of VRP.

The Vehicle Routing Problem covers both exact and heuristic methods developed for the VRP and some of its main variants, emphasizing the practical issues common to VRP. The objective of the problem is to design vehicle routes to meet certain requirements such as:

> ➢ minimising number of vehicles used
> ➢ minimising total distance travelled
> ➢ minimizing combination of number of vehicles used and total distance travelled

Here the cost of vehicles in vehicle fleet is regarded as fixed cost so that the first objective corresponds to minimisation of fixed costs, second objective corresponds to minimisation of running costs while third objective corresponds to minimisation of total cost.

Vehicle routing problem as encountered in practice involves many restrictions on routes that delivery vehicles can follow and consider some more common restrictions. The restrictions can be classified to certain extent as relating either to vehicles or to customers. In any particular case not all of these restrictions apply, but thinking generically about the problem is required to list all restrictions that can potentially apply.

Vehicles

➤ Each vehicle has a limit on goods carried.

➤ Each vehicle has total working time from departure to arrival back at depot.

➤ Each vehicle has time period in which it must leave the depot to ensure space availability for incoming vehicles to resupply the depot.

➤ Each vehicle has number of time periods during which it does nothing.

➤ Each vehicle has cost associated with its use for deliveries.

Customers

➤ Each customer has certain quantity that to be delivered, but there are operations involving collections and mix of collections and deliveries.

➤ Each customer has number of time periods during which delivery can occur. For example a customer might be prepared only to accept delivery between 10.30 and 11.30 or between 14.00 and 16.15. These two periods of time are the time windows for the customer. Time windows are convenient to customers as they know when delivery is likely to occur and they can schedule deliveries to suit the work pattern of their staff. Time windows are inconvenient to delivery companies as they limit their flexibility.

➤ Each customer has an associated visit time.

➤ Each customer has a set of vehicles which cannot be used for delivery.

➤ Each customer has a priority for delivery. Typically this might happen due to driver/vehicle unavailability or due to poor weather conditions dramatically reducing vehicle speeds

➤ Each customer may accept split visits or not.

Other Factors

➤ Multiple trips by same vehicle on single day, where vehicle returns to depot and goes out again

➤ Trips by same vehicle longer than one day

➤ Compartmentalised vehicles with many different types of product to deliver. Petrol (gasoline) tankers are often compartmentalised, as are food delivery vehicles

➤ More than one depot, where vehicles can start/visit/end at different depots

6.5.2 Solution Methods for VRP

There are certain commonly used techniques for solving Vehicle Routing Problems. Among all, heuristics and metaheuristics methods are commonly applied as no exact algorithm can be guaranteed to find optimal tours in reasonable computing time when number of cities are large.

Exact Approaches

As the name suggests, this approach proposes to compute every possible solution until one of the bests is reached.

> ➢ Branch and bound
> ➢ Branch and cut

Heuristics

Heuristic methods perform a relatively limited exploration of the search space and typically produce good quality solutions within modest computing times. The heuristic approach is based on Clarke-Wright algorithm that solves open version of capacitated vehicle routing problem where vehicles are not required to return to depot after completing service. The proposed CW has been presented in four procedures of Clarke-Wright formula, open-route construction, two-phase selection, and route post improvement.

The fundamental structure of VRPB route consists of three parts:

> ➢ Hamiltonian path from DC through all delivery points and ending at delivery interface point.
> ➢ Interface link between delivery and pickup customers.
> ➢ Hamiltonian path from DC through all pickup points and terminating at pickup interface point.

The set of delivery customers on delivery path comprises sector of plane anchored at DC. A similar sector is defined by the set of pickup customers on the pickup path. Jacobs-Blecha showed that the best savings from backhauling can be attained by minimizing the angles of the delivery and pickup sectors as well as the angle between the delivery and pickup sectors, as shown in figure. This property will be exploited in the initialization phase of the LHBH algorithm.

The algorithm LHBH is based on Generalized Assignment Problem, and is similar to Fisher and Jaikumar GAP heuristic for VRP. However, this method differs most from Fisher and Jaikumar's approach in two respects. LHBH employs a fresh, new method for executing the process known as route seeding (Figure 6.2).

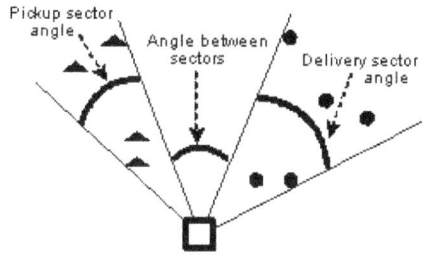

Figure 6.2: Important Angles in the VRPB.

The algorithm comprises three phases:

1. initialization,
2. clustering,
3. sequencing.

In initialization phase, an initial solution is obtained and costs are estimated for solving clustering problem. In phase (2), costs from (1) are used to solve Generalized Assignment Problem, which allocates delivery and pickup customers to set minimal cost routes. Phase (3) is concerned with solving the TSP for each cluster formed in (2)

Initialization

The idea of initialization phase is to provide a set of approximate locations for route clusters. As the concept of route locations is nebulous, spatial perspective of problem shows that customer proximity plays a strong part. From route structure, locations are sought that tends to minimize the route sector angles and angles among delivery and pickup sectors. For fiinding route clusters, the multifacility location model is used, viewing the demand points as existing facilities and cluster locations as new facilities that gives fundamental new way of initializing Fisher and Jaikumar approach to vehicle routing problems.

In determining location of route clusters, initialization phase starts by uniformly locating radials around DC to show every VRPB routes. In real-time, dispatcher placed these initial radials as per known streets and roads. The location and distance among all pickup with delivery customers are computed using distance metric based on angles, ring-radial metric as per route structure analysis. The facility location problem is solved iteratively as capacitated location-

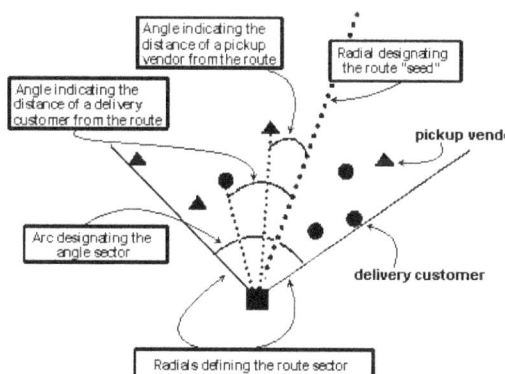

Figure 6.3: Ring-Radial Metric for a Single VRPB Route.

allocation problem with single sourcing constraints relaxed. This allows cluster determination to be influenced both by spatial distribution and customer demand. This approach has the advantage of achieving good seed locations for both uniformly and clustered distributions of customer facilities.

The ring-radial metric is distance metric used in facility location model as shown in figure that shows the metric where θj be angle of demand point j in polar coordinate plane and θi be the angle of radial showing route cluster I where θj is smallest angle that subtends at DC by these radials, hence, $\theta_{ij} = \min\{|\theta i - \theta j| \; 2\pi - |\theta i - \theta j|\}$

In solving facility location allocation problem, θ_{ij} is used to measure the distance between a customer and route location. The solution to the problem shows a set of angles which shows the radials along which each VRPB route is located. This set of radials serves as route seeds for the initial VRPB route locations. As observed, it is possible for customer allocations to get in this initialization phase to be infeasible with respect to single sourcing. Only route locations are applied in next phase of solution procedure, and any such infeasibility is eliminated when the clustering is completed.

Clustering

In clustering, there are two tasks to be performed:

1. determining cost of assigning a customer to route
2. use costs to make route assignments by solving Generalized Assignment Problem

After establishing seed radials, a route primitive is generated by choosing points for route that are near the seed radial. For each route, linehaul points are sequenced by increasing distance to the distribution center and backhaul points by decreasing distance. Any point which is within 10 degrees of more than one seed radial is not placed in either primitive, and left unassigned. This assignment of points results in a polygonal route primitive from which the GAP costs can be determined, such a set of route primitives is shown in Figure 6.4.

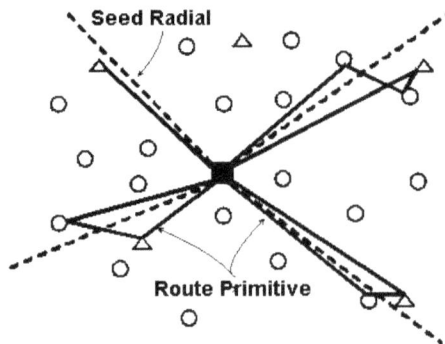

Figure 6.4: Route Primitives for VRPB.

After the route primitives is obtained, Euclidean distance metric is applied in remainder of the clustering phase. For problems where customers are randomly located in areas around DC, routes should be determined that fits the model of minimal sector angles. Here, Euclidean distance is a better estimator of nearness than the ring-radial metric as customer locations are naturally clustered, so ring-radial metric is better.

The method assigns points to routes in a sequential manner based on a computation of regret to be experienced by waiting until later to make the assignment. Now the cost of assigning each remaining point i to route k is estimated as minimum insertion cost of point i in links of the primitive for route k. As savings regret heuristic is sequential in it assignments, every time a point is assigned to route, primitive for that route grows and insertion costs for points to be assigned to route that change.

Constructive Methods

Gradually build a feasible solution while keeping an eye on solution cost, but do not contain an improvement phase.

➤ Savings: Clark and Wright

➤ Matching Based

➤ Multi-route Improvement Heuristics

Improvement/Exchange: Perhaps the best known method is the r-opt algorithm of Lin and Kernighan. Other exchange procedures exchange customers between routes, instead of within routes. Such methods can easily be applied to a given solution for VRPB by taking into account the precedence relationship of deliveries before pickups whenever an exchange is considered.

2-Phase Algorithm

The problem is decomposed into its two natural components: (1) clustering of vertices into feasible routes and (2) actual route construction, with possible feedback loops between the two stages.

➤ Cluster-First, Route-Second Algorithms

➤ Route-First, Cluster-Second Algorithms

Cluster-first/Route-second: This strategy is shown by sweep algorithm of Gillett and Miller. The sweep approach extends to VRPB by truncating clusters when either linehaul or backhaul capacity exceeds.

Route-first/Cluster-second: Extension of this approach to VRPB can is done by solving a Traveling Salesman Problem for delivery points and then solving TSP for pickup points. Each of large tours can be broken into individual delivery and pickup routes that can be patched together forming linehaul-backhaul routes.

➤ Metaheuristics

➤ Ant Algorithms

- ➤ Constraint Programming
- ➤ Deterministic Annealing
- ➤ Genetic Algorithms
- ➤ Simulated Annealing
- ➤ Tabu Search

6.6 Determination of Size and Schedules for Transportation Fleets

Fleet size calculation for intercity passenger service is an important item in the intercity transportation planning. The intercity transportation planners in companies face problems in determining the optimum number of vehicles that can operate a certain service.

Fleet size determination based on scientific bases is one of the main issues that can contribute in solving many problems. The importance of this issue can be reflected as follows:

- ➤ To help the private sector to conduct the feasibility study of its transportation project.
- ➤ To alleviate problems resulted from terminal congestion.
- ➤ To best utilize the current fleet size of the intercity bus operators.

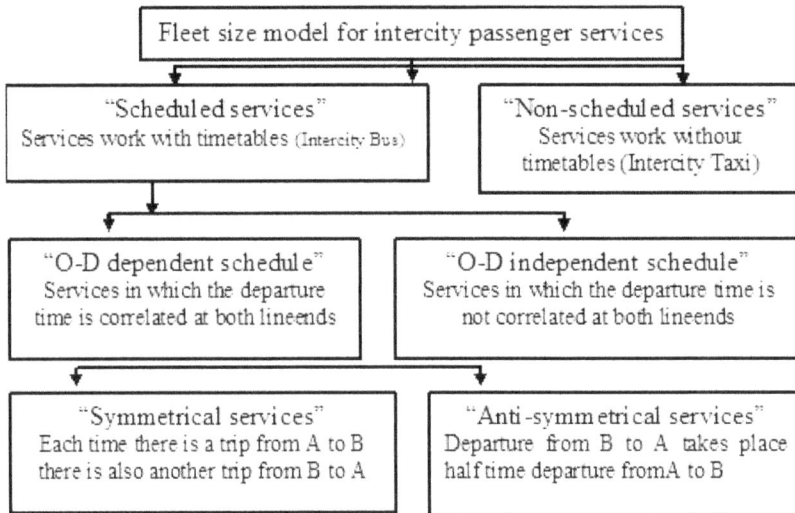

Figure 6.5: Intercity Services Classification.

There are different types of intercity public services based on the departure pattern at both trip ends. As discussed, under certain conditions, departure time at both ends is dependent. However, these services may operate as a day service, continuous service and night Service (Figure 6.5). A simple model was developed for each of the afore-mentioned three services. These models were developed under the condition that all services are symmetrical. The day service based model takes the following formulae:

$$N_{fr} = 2 * Int\left(\frac{T_b + T_t}{T_s}\right)$$

$$N_{fr} \leq 2 N_{sr}$$

$$\leq 2\left(\frac{T}{T_s} + 1\right) = 2\,int\left(\frac{T}{T_s} + 1\right)$$

Where:

N_{fr} = the fleet size required to operate the regular service

N_{sr} = number of daily regular services

$int\,(x)$ = the smaller integer value of the real number (x).

T_s = time between two successive departures in the regular service in hours;

N_{sd} = total number of departures per day;

N_{sp} = extra number of departures in the two peak periods

T_n = interval time of the night period in hours;

T_b = the journey time or the block time in hours;

T = 24-T_n –T_b = total working hours for the service;

T_t = terminal time

T_p = interval time for one peak period.

It is observed that the fleet size required to operate the regular service in symmetrical service is the smaller value as given by to equations above. During the two equal peak periods, it was assumed that number of departures would be doubled, so the extra departures for peak periods have the same symmetrical time but can be shifted by $0.5T_s$ from regular service. It shows that the time between each two successive departures in peak period is $0.5T_s$ will be:

$$N_{sp} = 2 * int\left(\frac{T_p (N_{sd} - 1)}{24 + 2T_p - T_n - T_b}\right)$$

Here the extra number of vehicles required for each peak period can be calculated as $0.5N_{sp}$ departures over a time $(0.5N_{sp}-1)T_s$, which resulting in following equation:

$$N_{fp} = 2 * Int\left(\frac{T_b + T_t}{T_s}\right)$$

Where

N_{fp} = fleet size required to operate the peak periods.

The fleet size required for peak period must not exceed the number of departures during the peak period, *i.e.,*

$$N_{fp} \le 2N_{sp}$$

Now the number of vehicles required to operate extra number of departures during two peak periods will be of smaller value as given by above equations. Hence, the fleet size required to operate the symmetrical service with two peak periods will be:

$$n_{fleet} = N_{fr} + N_{fp}$$

But, fleet size determined by the model tends to be even number that leads extra cost in small fleet sizes, so the model assumed that the journey time will be same in two directions all over the day, that perhaps is not correct. At the same time, main variables affecting fleet size like vehicle capacity, maximum and minimum headway with demand requirement will not taken into account. Hence, the model analyzed peak periods from two equal, in time, peak periods. It was assumed that both the demand and the frequency were doubled during the two peak periods.

A simple model to calculate the intercity fleet size for anti-symmetric service has been also developed which was based on three service types; day, continuous and night. In order to overcome limitations of symmetrical service-based model, model is extended to allow for anti-symmetrical services. This assumption led to a fleet size containing an odd number of vehicles. The anti-symmetrical service model was built on the same assumptions of the symmetrical one.

The fleet size required to operate the anti-symmetrical services was given by

$$Nfr = 2 * \text{int}\left(\frac{Tb + Tt}{Ta} - \frac{1}{2}\right) + 1$$

$$N_{fr} = 2 * N_{sr} = 2 * (N_{sd} - N_{sp})$$

In symmetrical services model, it is seen that frequency in peak hour will get doubled, so the extra number of services is similar to:

$$Nsp = 2 * \text{int}\left(\frac{Tp(2Nsd - 1)}{2(24 - Tn - Tb + 2Tp)}\right)$$

The peak period was considered as another anti-symmetrical service. Then, the number of vehicles required to operate the peak period is the smaller value of:

$$Nfp = 2 * \text{int}\left(\frac{Tb + Tt}{Ta} - \frac{1}{2}\right) + 1 \text{ and } N_{fp} = Nsp$$

The calculation of fleet size model for scheduled services with no relation between departures at both ends was prepared.

The fleet size for a given demand is affected by many factors. These are:

- ➤ Trip characteristics factors such as trip time.
- ➤ Vehicle characteristics factors such as seating capacity.
- ➤ Operation policy related factors such as:
  - demand based headway
  - maximum headway
  - minimum terminal time
  - maximum daily travel distance

Demand Based Headway

The headway can be calculated based on the total daily passengers traveled by the service and the seating capacity of the fleet as follows:

Total departures in service time:
Nsr = Total daily passengers per direction (P)/Load factor (L)*Seating capacity (S)
Nsr = P/LS

The load factor (L) is percentage of passengers (P) in bus to seating capacity (s). This factor varies between 0.56 and .95. However, for design purpose, it can be assumed that L = 1 in direction of maximum demand and less than one in other direction. The headway can be calculated as follows:

Headway= (Service time)/(Total departures in service time-1)
=T/Nsr-1

The headway calculated based on the travel demand can not be considered for the service design without taking into consideration some other factors that determine both the maximum and minimum headway. Maximum and minimum headway should be first calculated to check that the travel demand based headway satisfies both limits.

Maximum Headway Calculation

The maximum waiting time that the passenger can wait before deciding to use another service is the main factor affecting the maximum allowable headway. This factor has been a part of the data collection program as shown in Figure 6.6.

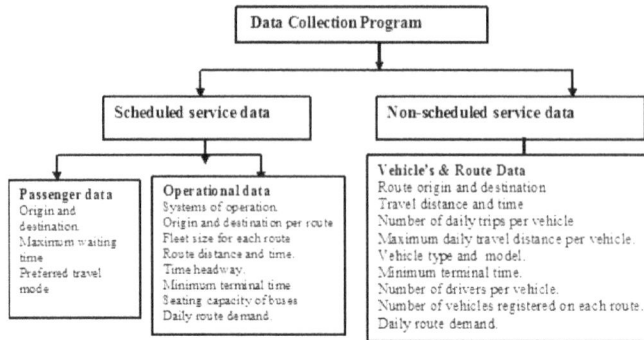

Figure 6.6: Data Collection Items.

The following steps have been carried out for maximum waiting time calculation:

A random sample of about 472 passengers was interviewed. For this sample, maximum waiting times before deciding to use another mode of transport were collected.

For more accurate results, journey time for samples was divided into three categories as per bus journey time. The percentage of maximum waiting time to trip time was calculated for each category as shown in Table 6.1.

Table 6.1: Maximum Waiting Time.

Item	Journey Time Category (hours)		
	< 2	from 2 to 5	> 5
Sample size	292	85	95
Average percentage of waiting time to the journey time	0.415	0.217	0.153
Standard deviation	0.333	0.0995	0.083

The table reveals that maximum percentage of waiting time decreases as the journey time increases and vice verse.

Minimum Terminal Time

Once the data is collected, it is analyzed to find minimum required terminal time. As observed, that there is no time required for maintenance as it is done at the end of service time every day, however, some time might be needed for clearing. The time spent for passenger loading and unloading is the only criterion that governs the minimum headway time. The loading and unloading time depends on the seating capacity and the service time taken by each passenger. Mathematically, this can be expressed as follows:

Minimum terminal time = S*t1 + S*t2

Where:

S = seating capacity

t1, t2 = service time required for one passenger for alighting and boarding,

The values of service time for alighting and boarding are usually between 4 and 6 seconds and 6 and 8 seconds per passenger, assuming that passenger has considerable amount of baggage. As per data and interviews it were concluded that terminal time for bus should not be less than 15 minutes as per desired conditions.

Maximum Daily Travel Distance per Bus

The daily travel distance is controlled to avoid over use, which leads to a shortage. It was found that the intercity transportation replace buses after 10 working years. After that, buses become less efficient and more costly. The annual and daily travel distance are depicted in Table 6.2.

Table 6.2: Annual and Daily Travel Distance

Service type	Average Annual Km	Average Daily Km
Suppe Deluxe-Air conditioned	211000	700
Deluxe-Air conditioned	178000	600
Express	126400	450
Common	67400	250

6.7 Synchronization of Signalized Interactions

In Hodgkin-Huxley model of neurons, the phase diagram of developing single-layer neural network were investigated where network comprises of two weakly coupled neural layers. These networks are noise driven and learn through spike-timing-dependent plasticity (STDP) or inverse STDP rules.

In neural network model and numerical method for investigating interaction among two coupled neural layers each having 50 neurons where activities of neurons in the network were modeled as per Hodgkin-Huxley neuron model theory. For activity-dependent development of neural networks, spike-timing-dependent plasticity (STDP) and inverse STDP were considered for the learning of synapses. In absence of an external input, in Section Neural Networks in a Noisy Environment, synaptic noise is modeled in noisy neural network to study intrinsic dynamic behaviors of network. In Section Simulating Coupled Neural Networks, various coupled neural networks are framed with possible types of neuron distribution on each layer.

The Neuron Model

A mathematical model explains the ionic mechanisms that underlies the initiation and propagation of action potentials in squid giant axon. In this model, each component of an excitable cell is treated as an electrical element. The lipid bilayer is represented as a capacitance (C_m). Voltage-gated ion channels and leak channels are represented by electrical conductance (g_{Na}, g_K, and g_L denote the maximum conductance per surface area of the sodium, potassium and leak currents). Finally, ion pumps are represented by current sources (I). Explicitly, the dynamics of neurons is described by the following equations:

$$CmdVi/dt=gNam3ihi(V_{Na}-V_i)+gKn4i(V_K-V_i) + gL(V_{rest}-V_i)+I^{syn}i(t),$$

$$dmi/dt=(1-mi) \times 25 - Vi/10[\exp (25-Vi/10)-1] - mi[4 \cdot \exp (-Vi/18)]$$

$$dni/dt=(1-ni) \cdot 1-0.1 \cdot Vi/10 [\exp (10-Vi/10)-1] - ni[0.125 \cdot \exp (-Vi/80)],$$

$$dhi/dt=(1-hi) \cdot 0.07 \cdot \exp (-Vi/20) - hi/\exp (30-Vi/10)+1$$

where a set of four time-dependent variables (V_i, m_i, n_i, h_i) were used to describe the activity of i-th neuron. Here V_i is the membrane potential, m_i and h_i are the activation and inactivation variables of the sodium current, and n_i is the activation variable of the potassium current. V_{Na}, V_k and V_{rest} are the corresponding reversal potentials. Typical values of the parameters were chosen as C_m = 1.0 μF/cm^2, g_{Na} = 120.0 mS/cm^2, g_K = 36.0 mS/cm^2, g_L = 0.3 mS/cm^2, V_{Na} = 115.0 mV, V_k = −12.0 mV, and V_{rest} = 10.6 mV. $Isyni(t)Iisyn(t)$ is the total synaptic current, a sum of output currents, $Ioutj(t-T0)Ijout(t-T0)$, from connected neurons in the network with a synaptic strength w_{ij} and a signal delay time T_0. Explicitly, the total synaptic current is expressed as $Isyni(t)=\sum jwij \cdot Ioutj(t-T0)Iisyn(t)=\sum jwij \cdot Ijout(t-T0)$. For simplicity we approximated the output current $I_j^{out}(t)$ as a step function with a duration 0.1 ms and an amplitude $Imax \cdot \{1+\exp[-0.002 \cdot Vpeakj(t)]\}-1Imax \cdot \{1+exp[-0.002 \cdot Vjpeak(t)]\}-1$, where $V_j^{peak}(t)$ is the peak value of the AP of j-th neuron at time t and I_{max} is the maximum output current from a neuron. Typical values of I_{max} and T_0 used in our simulations are 25 nA/cm^2 and 9 ms, respectively.

Synaptic Plasticity

In neural systems, a synapse between two neurons can change its strength in response to either use or disuse of transmission over synaptic pathways. Earlier, Hebbian learning rule was suggested that synaptic strength could increase if presynaptic neuron repeatedly and persistently stimulates the postsynaptic neuron to generate APs. More recent experiments have observed a spike-timing-dependent synaptic plasticity (STDP): repeated presynaptic spike arrival a few milliseconds before postsynaptic action potentials leads in many synapse types to long-term potentiation (LTP) of the synapses, whereas repeated spike arrival after postsynaptic spikes leads to long-term depression (LTD) of the same synapse. Previous experiments also demonstrated that postsynaptic APs are initiated in the axon and then propagate back into the dendritic arbor of neocortical pyramidal neurons, evoking an activity dependent dendritic Ca^{2+} influx that could be a signal to induce modifications at the dendritic synapses that were active around the time of AP initiation. Hence, synaptic efficacy can be regulated as per precise timing of postsynaptic APs relative to excitatory postsynaptic potentials. The characteristic time intervals for synaptic modifications are found to be 17 ms for facilitation and 34 ms for depression for layer 5 pyramidal neurons in somatosensory cortex.

7 | **Project Scheduling with Resource Constraints**

7.1 Introduction

Scheduling and sequencing is related to optimum allocation of scarce resources over time. It deals with defining which activities are to be performed at a particular time. Sequencing concerns the ordering in which the activities have to be performed. The allocation of scarce resources over time has been the subject of extensive research since the early days of operations research in way back in 1950s. The result is vast and not easy and a considerable gap between scheduling theory and shop floor practice. Practitioners blame scheduling theoreticians to spend scarce research money for studying toy problems such as sequencing a set of simultaneously available unordered jobs with known durations on a never failing machine in order to optimize irrelevant objective functions.

Theoreticians blame practitioners for their ignorance about the recent development, their reluctance in applying useful theory, or their over-enthusiasm in applying scheduling procedures miles away from their natural field of application. Despite this mutual 'interest', the major issues largely remain unresolved in practice and scheduling and sequencing problems remain the subject of intensive research.

It is noted that activities are commonly called as jobs, and it is usually assumed that a job is processed by a single machine at a time. The processing of job on a machine is called an operations. The machine environment is quite diverse. In a single machine environment, each job has only one operation while in parallel machine environment, each job requires just one operation and that operation may be performed on any of the machines. When the machines are identical, the processing time of a job is the same on all machines. When the machines are uniform, the processing time varies as a function of given reference speed. When the machines are unrelated, the processing time of job again varies, but now in a completely arbitrary fashion. In multistage production, a job consists of number of operations.

Technological procedures constraints demand that each job should be processed through the machines in a particular order. For general jobs problems there are no restrictions upon the form of the technological constraints. When all the jobs share the same processing order we have a flow-shop problem. In the special case of an open shop, each job has to be processed on each machine, but there is no particular order to follow. In open shops the schedule determines not only the order in which machines process the jobs, but also that in which the jobs pass between machines. Jobs are characterized by a ready time which denotes the time at which the job becomes available for processing. The time by which the job should be finished is called the due date. It is possible to consider situations were jobs may be splitted or not. Each operation takes a certain length of time, the processing time, to be performed. In addition, operations may suffer from sequence dependent set up times.

The most common approaches to project scheduling are Critical Path Method (CPM) and Project Evaluation and Review Technique (PERT). Although CPM is strictly deterministic and with PERT there is an uncertainty in the activity time estimates, the methods within these techniques are quite similar. Both are computer-oriented, define arrow network diagrams, and define the concept of a critical path. For these reasons, PERT and CPM are referred to jointly in this paper as PERT/CPM.

7.2 Project Scheduling

Project scheduling is an act of framing a timetable for each project activity which differs in complexity due to presence of renewable resources with limited availability. There are three important aspects of scheduling such as:

> ➢ Sequencing: scheduling with unlimited resources

> ➢ Scheduling: scheduling within limited resource constraints

> ➢ Scheduling objectives

Project scheduling involves construction of activity timetable, such as determination of a start and finish time for each project activity, respecting the precedence relations and the limited availability of the renewable resources, while optimizing a predefined scheduling objective.

Sequencing

In the absence of renewable resource constraints, project scheduling boils down to activity sequencing by putting each individual activity as-soon-as-possible in the timetable, respecting the precedence relations, resulting in an earliest start schedule. Consequently, in this scheduling approach, it is implicitly assumed that the minimization of the total project lead time is the scheduling objective.

Two well-known techniques that rely on a straightforward activity sequencing approach are the PERT and CPM scheduling techniques. More information can be found at "The Program Evaluation and Review Technique (PERT): Incorporating activity time variability in a project schedule and Critical Path Method (CPM): Incorporating activity time/cost trade-offs in a project schedule".

Scheduling

The presence of renewable resources, constrained by their limited periodic availability, leads to a complexity increase during the construction of a project schedule. Due to the limited availability of resources, the straightforward activity sequencing approach often leads to so-called resource conflicts. These conflicts result from over-allocations of renewable resources when activities scheduled in parallel require more resources than available. In order to solve these resource conflicts, activities need to be shifted further in time to periods where resources are still available for the activities. The aim of this scheduling approach is to create a so-called resource-feasible schedule and is often a complex task. Moreover, the construction of such a resource-feasible project schedule requires a scheduling objective that needs to be optimized.

Scheduling Objectives

A scheduling objective is the objective one aims to reach while constructing a resource-feasible project schedule. Although time is often considered as the dominant scheduling objective, other objectives are often crucial from a practical point-of-view. A non-exhaustive list of possible objectives is given along the following lines:

> ➢ Time: minimize the total duration of the project
>
> ➢ Net present value: maximize the discounted cash flows of project activities
>
> ➢ Work continuity: avoid idle time of bottleneck resources in a project
>
> ➢ Leveling: avoid resource jumps but try to balance the use of resources
>
> ➢ Others: Many other scheduling objectives can occur and are often project specific. Moreover, it is logical that in practical environments, a combination of objectives is strived for

7.3 Network Problems

For finding heuristic procedures selected, test problems from Patterson's benchmark problem set with optimal solutions and a heuristic solution were used. Such problems shows an accumulation of all readily available multiple resource problems existing in literature. Optimal durations included with problem set were obtained by Stinson Branch and Bound. The number of activities included in the test problems varies between 7 and 51, with the number of resource types required per activity varying between one and three with only four networks having less than three resources.

As per the study, ten simple priority-rule based heuristics are used to determine priorities for activities competing for constrained resources. They vary from simple single attribute heuristic to simple multiple attribute heuristic and include rules found effective in previous literature. The heuristics used to determine the priorities and the explanation of their coding is:

1. Longest Activity First (LAF): Priority given to the activity with the longest activity.

2. Shortest Job First (SJF): Priority given to the activity with the shortest duration.

3. First Come First Served (FCFS): Priority given to the activity with the lowest activity number.

4. Latest Finish Time (LFT): Priority given to the activity with the earliest PERT/CPM calculated late finish time.

5. Minimum Early Finish (MEF): Priority given to the activity with the earliest PERT/CPM calculated early finish time.

6. Minimum Slack First (MSF): Priority given to the activity with the least PERT/CPM calculated slack time.

7. Maximum Slack First (Max SF): Priority given to the activity with the greatest PERT/CPM calculated slack time.

8. ACTIM: Priority given to the activity with the maximum ACTIM value. The ACTIM value of an activity is calculated as the maximum time that the activity controls through the network on any one path.

9. ACTRES: Priority given to activity with the maximum ACTRES value. The ACTRES value is calculated by multiplying each activity's time by the sum of its resources and then finding the maximum ACTRES that an activity controls through the network on any one path.

10. Resources Over Time (ROT): Priority given to the activity with the maximum ROT value. The ROT value is calculated by dividing the sum of each activity's resources by the duration of the activity and then finding the maximum ROT that an activity controls through the network on any one path.

The heuristics are used with the serial approach, with activity priority being determined during the scheduling procedure, but based on PERT/CPM calculations obtained at the beginning of the scheduling procedure. For all heuristics, ties are broken by the lowest activity number first and then once an activity is started it is not interrupted.

7.4 Critical Path Method (CPM)

It is a wise and generally accepted management principle to put a focus on the constraining or limiting factor of a system that determines the system's goal. In project management and scheduling, the scheduling objective is the objective during the construction of a project baseline schedule.

It is assumed that the scheduling objective is the minimization of the total project duration. A distinction is made between the construction of a schedule with and without renewable resource constraints, as summarized along the following lines:

> ➤ The critical path: no renewable resources limit the scheduling degrees of freedom

➤ The critical chain: renewable resources limit the scheduling degrees of freedom

7.4.1 Critical Path

Figure below shows a project network with certain activities and finish-start precedence relations between them. Each number above the node denotes the estimated duration of the activity. In the PERT/CPM scheduling approach, projects are not subject to limited renewable resources, and hence, the construction of a baseline schedule boils down to sequencing all activities according to their precedence relations.

The minimal duration of the project is determined by the length of the critical path, which is a sequence of activities in the project network, as shown in figure 1 in red and bold. The attentive reader has noticed that a second path (1 - 2 - 5 - 8 - 10 - 11) is also equally critical since it has the same total duration of 22 time units. Indeed, in a network, multiple critical paths can occur (Figure 7.1).

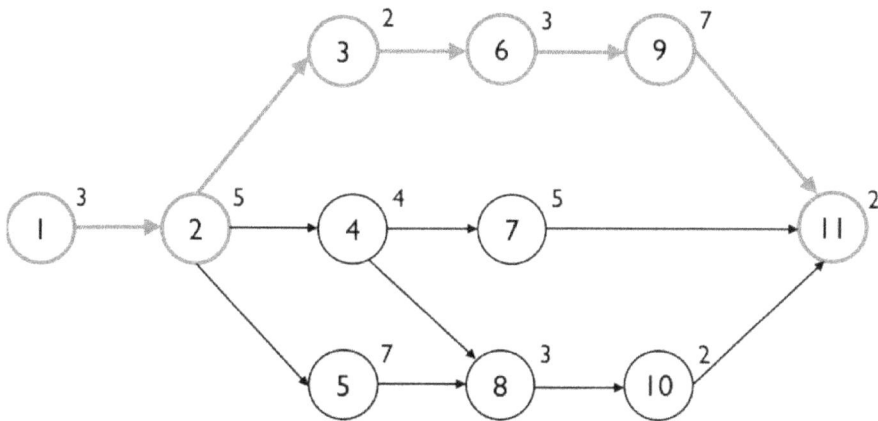

Figure 7.1: Project Network with Critical Path.

7.4.2 Critical Chain

Scheduling projects within the presence of resources with a limited availability might lead to resource conflicts. In below figure, it is assumed that the availability of the resource is restricted to six units for all periods of the project. The earliest start schedule displayed at the top of the figure shows a resource conflict between period 8 and 13 since the total use of the renewable resource by activities 3, 4, 5, 6 and 7 exceeds its limited availability of 6. In order to solve this resource conflict, activities need to be shifted further in time to time periods where resources are still available. The way these resource conflicts have to be solved depends on the scheduling objective. The bottom schedule of below figure shows a resource-feasible schedule with a minimal project duration of 24 time periods as a result of shifts in activities 5 and 6 (Figure 7.2).

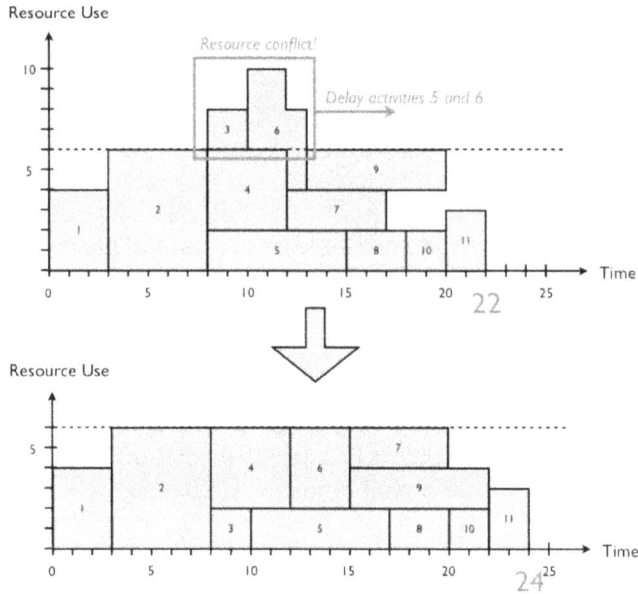

Figure 7.2: Resource Profiles for Critical Path Schedule.

A comparison between the resource-unconstrained critical path schedule (length = 22) and the resource-feasible resource-constrained schedule (length = 24) leads to the following two conclusions:

➢ The critical path duration is always smaller than or equal to the resource-feasible project schedule duration. Indeed, since the limited availability of renewable resources puts an extra constraint on the degrees of freedom during the construction of a baseline schedule, the critical path length is a lower bound on the length of the resource-feasible schedule duration.

➢ In analogy to the critical path, which determines the subset of project activities that are responsible for the length of the project schedule, a similar concept, known as the critical chain, holds for a resource feasible schedule. However, while the critical path is the result of a simple sequencing approach, the critical chain depends on the scheduling objective and hence on the choice of activities that have been delayed. In the example, activities 5 and 6 have been delayed to solve the resource conflict, but another scheduling objective could result in other activities to delay and hence, another critical chain. The critical chain in the example is equal to the activity sequence 1 - 2 - 4 - 6 - 9 - 11 and consists of precedence relations as well as, unlike the critical path, resource relations. Figure 3 shows the resource link between activities 4 and 6 that has been added to the project network (the number below the node is used to refer to the resource demand). The extra resource link is added to prevent that activities 4 and 6 will be scheduled simultaneously (Figure 7.3).

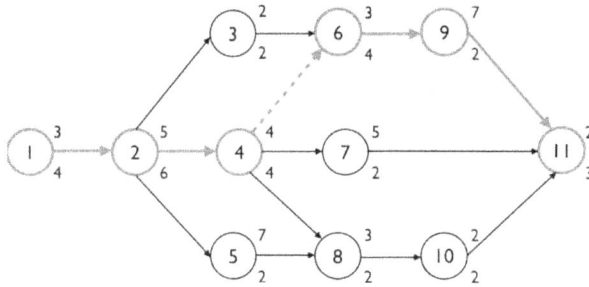

Figure 7.3: Project Network with Critical Chain with 24 Time Periods.

Example

Consider the network diagram shown below and find the total paths, critical path and float for each path.

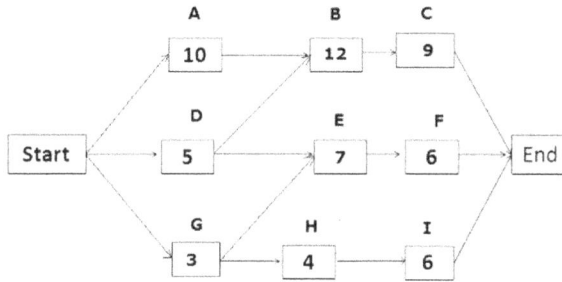

Network Flow

The above network diagram has five paths; the paths and their duration are as follows:

1. Start -> A -> B -> C-> End, duration: 31 days.
2. Start ->D -> E ->F -> End, duration: 18 days.
3. Start -> D -> B -> C -> End, duration: 26 days.
4. Start -> G ->H ->I -> End, duration: 13 days.
5. Start -> G -> E ->F -> End, duration: 16 days.

Since the duration of the first path is the longest, it is the critical path. The float on the critical path is zero. The float for the second path "Start ->D -> E ->F -> End" = duration of the critical path – duration of the path "Start ->D -> E ->F -> End"

= 31 – 18 = 13
Hence, the float for the second path is 13 days.
Using the same process, we can calculate the float for other paths as well.
Float for the third path = 31 – 26 = 5 days.
Float for the fourth path = 31 – 13 = 18 days.
Float for the fifth path = 31 – 16 = 15 days.

Calculating Early Start (ES), Early Finish (EF), Late Start (LS), and Late Finish (LF)

We have identified the critical path, and the duration of the other paths, it's time to move on to more advanced calculations, Early Start, Early Finish, Late Start and Late Finish.

Calculating Early Start (ES) and Early Finish (EF)

For finding Early Start and Early Finish dates, use forward pass by starting from the beginning and proceed to the end. Early Start (ES) for first activity on any path will be 1, because no activity can be started before the first day. The start point for any activity or step along the path is the end point of the predecessor activity on the path plus one.

The formula used for calculating Early Start and Early Finish dates.

> ➤ Early Start of the activity = Early Finish of predecessor activity + 1
> ➤ Early Finish of the activity = Activity duration + Early Start of activity − 1

Early Start and Early Finish Dates for the path Start -> A -> B -> C -> End

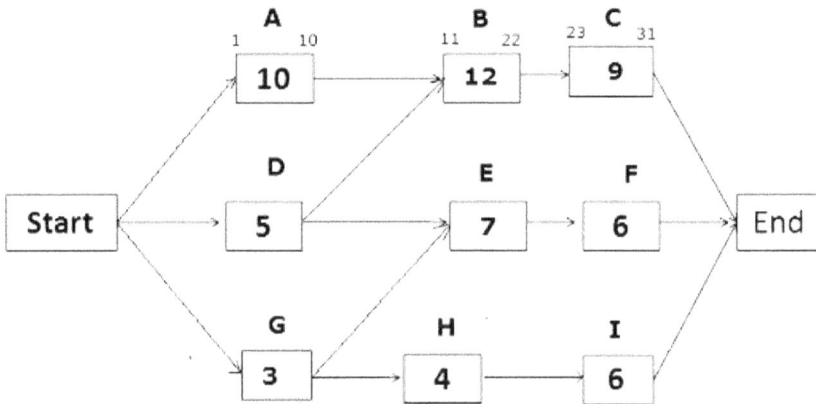

Early Start of activity A = 1 (Since this is the first activity of the path)
Early Finish of activity A = ES of activity A + activity duration − 1
= 1 + 10 − 1 = 10
Early Start of activity B = EF of predecessor activity + 1
= 10 + 1 = 11
Early Finish of activity B = ES of activity B + activity duration − 1
= 11 + 12 − 1 = 22
Early Start of activity C = EF of predecessor activity + 1
= 22 + 1 = 23
Early Finish of activity C = ES of activity C + activity duration − 1
= 23 + 9 − 1 = 31

Early Start and Early Finish Dates for the path Start -> D -> E -> F -> End

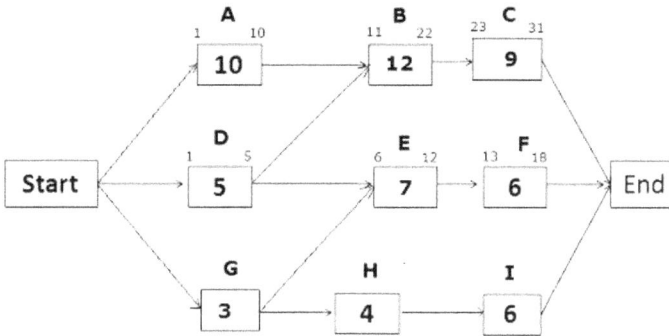

Early Start of activity D = 1 (Since this is the first activity of the path)

Early Finish of activity D = 1 + 5 – 1 = 5

Early Start of activity E = EF of predecessor activity + 1

Since the Activity E has two predecessor activities, which one will you select? You will select the activity with the greater Early Finish date. Early Finish of activity D is 5, and Early Finish of activity G is 3.

Therefore, we will select the Early Finish of activity D to find the Early Start of activity E.

Early Start of activity E = EF of predecessor activity + 1
= 5 + 1 = 6
Early Finish of activity E = 6 + 7 – 1 = 12
Early Start of activity F = 12 + 1 = 13
Early Finish of activity F = 13 + 6 -1 = 18
Early Start and Early Finish Dates for the path Start -> G -> H -> I -> End

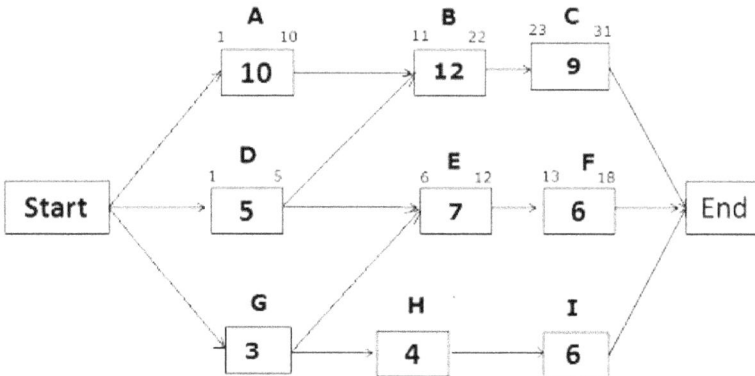

Early Start of activity G = 1 (Since this is the first activity of the path)
Early Finish of activity G = 1 + 3 – 1 = 3

Early Start of activity H = 3 + 1 = 4
Early Finish of activity H = 4 + 4 – 1 = 7

Early Start of activity I = 7 +1 = 8
Early Finish of activity I = 8 + 6 – 1 = 13

Calculating Late Start (LS) and Late Finish (LF)

After calculating Early Start and Early Finish dates of all activities, now calculate the Late Start and Late Finish dates. Late Finish of the last activity in any path will be the same as the Last Finish of the last activity on the critical path, because you cannot continue any activity once the project is completed. The formula used for Late Start and Late Finish dates:

   ➢   Late Start of Activity = Late Finish of activity – activity duration + 1

   ➢   Late Finish of Activity = Late Start of successor activity – 1

To calculate the Late Start and Late Finish, we use backward pass; *i.e.* we will start from the last activity and move back towards the first activity.

Late Start and Late Finish Dates for the path Start -> A -> B -> C -> End

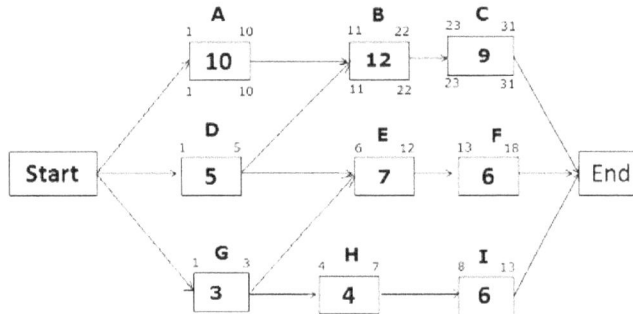

On a critical path, Early Start, and Early Finish dates will be the same as Late Start and Late Finish dates.

Late Start and Late Finish Dates for the path Start -> D -> E -> F -> End

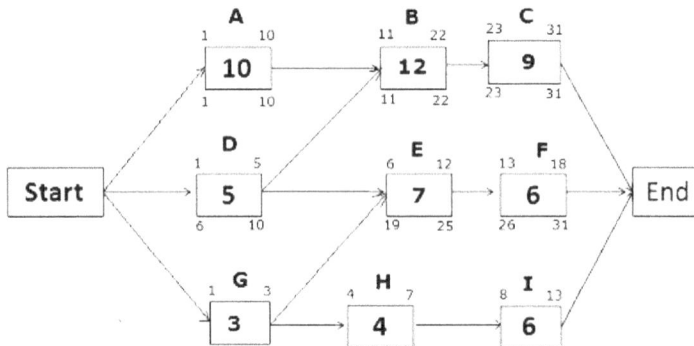

Late Finish of activity F = 31 (because you cannot allow any activity to cross the project completion date)

Late Start of activity F = LF of activity F – activity duration + 1
= 31 – 6 +1 = 26
Late Finish of activity E = LS of successor activity – 1
= LS of activity F – 1
= 26 – 1 = 25
Late Start of Activity E = LF of activity E – activity duration + 1
= 25 – 7 + 1 = 19
Late Finish of activity D = LS of successor activity – 1

After looking at network diagram, it is observed that activity D has two successor activities, B and E, so selecting activity with earlier(least) Late Start date. Here, Late Start of activity B is 11, and Late Start of activity E is 19. Hence, select activity B which has earlier Late Start date, so,
Late Finish of activity D = LS of activity B – 1
= 11 – 1 = 10
Late Start of Activity D = LF of activity D – activity duration + 1
= 10 – 5 + 1 = 6

Late Start and Late Finish Dates for the path Start -> G -> H -> I -> End

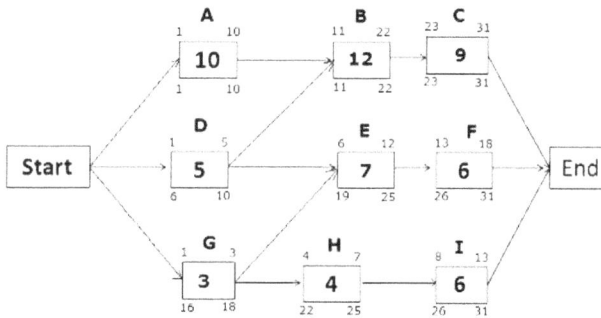

Late Finish of activity I = 31 (because you cannot allow any activity to cross the project completion date)

Late Start of activity I = 31 – 6 + 1 = 26
Late Finish of activity H = 26 – 1 = 25
Late Start of activity H = 25 – 4 + 1 = 22
Late Finish of Activity G = 19 – 1= 18
Now select the late start of activity E and not activity H as Late Start of activity E is earlier than the Late Start of activity H.
Late Start of activity G = 18 – 3 + 1
= 16

Calculating Free Float

The formula for Free Float is:
Free Float = ES of next activity – EF of current activity – 1

7.5 Project Evaluation and Review Technique (PERT)

PERT or Program Evaluation and Review Technique is a widely used method for planning and coordinating large-scale projects. It is basically a management planning and control tool which can be considered as a road map for a particular program or project in which all of the major elements have been completely identified, together with their corresponding interrelations'.

PERT charts are constructed from back to front as in many projects, end date is fixed and the contractor has front-end flexibility. The basic element of PERT-style planning is to find critical activities on which others depend. The technique is referred as PERT/CPM. PERT was developed way back in 1950s.

The main features of PERT analysis is network diagram which shows visual depiction of major project activities and sequence where they gets completed. Activities are defined as distinct steps toward completion of the project that consume either time or resources. The network diagram consists of arrows and nodes and can be organized using one of two different conventions. The arrows represent activities in the activity-on-arrow convention, while the nodes represent activities in the activity-on-node convention. For each activity, managers provide an estimate of the time required to complete it.

The sequence of activities leading from starting point of diagram to finishing point of diagram is called a path. The amount of time required to complete the work involved in any path can be figured by adding up the estimated times of all activities along that path. The path with the longest total time is then called the critical path, or CPM. The critical path is the most important part of the diagram for managers: it determines the completion date of the project. Delays in completing activities along the critical path necessitate an extension of the final deadline for the project. If a manager hopes to shorten the time required to complete the project, he or she must focus on finding ways to reduce the time involved in activities along the critical path.

The time estimates managers provide for the various activities comprising a project involve different degrees of certainty. When time estimates can be made with a high degree of certainty, they are called deterministic estimates. When they are subject to variation, they are called probabilistic estimates. In using the probabilistic approach, managers provide three estimates for each activity: an optimistic or best case estimate; a pessimistic or worst case estimate; and the most likely estimate. Statistical methods can be used to describe the extent of variability in these estimates, and thus the degree of uncertainty in the time provided for each activity. Computing the standard deviation of each path provides a probabilistic estimate of the time required to complete the overall project.

7.5.1 Analysis

For complex problems involving hundreds of activities, computers are used to create and analyze the project networks. The project information input into the computer includes the earliest start time for each activity, earliest finish time for each activity, latest start time for each activity, and latest finish time for each activity without delaying the project completion. From these values, a computer algorithm can determine the expected project duration and the activities located on the critical path. Solution of algorithm is easy for computer, but the resulting information will only be as good as the estimates originally made. Thus PERT depends on good estimates and sometimes inspired guesses.

PERT offers a number of advantages to managers. For example, it forces them to organize and quantify project information and provides them with a graphic display of the project. It also helps them to identify which activities are critical to the project completion time and should be watched closely, and which activities involve slack time and can be delayed without affecting the project completion time. The chief disadvantages of PERT lie in the nature of reality. Complex systems and plans, with many suppliers and channels of supply involved, sometimes make it difficult to predict precisely what will happen. The technique works best in well-understood engineering projects where sufficient experience exists to predict tasks accurately in advance.

The Program Evaluation and Review Technique (PERT) is a network model that allows for variations in activity completion times. In a PERT network model, each activity is represented by a line (or arc), and each milestone is shown by a node. A simple example is shown in Figure 7.4.

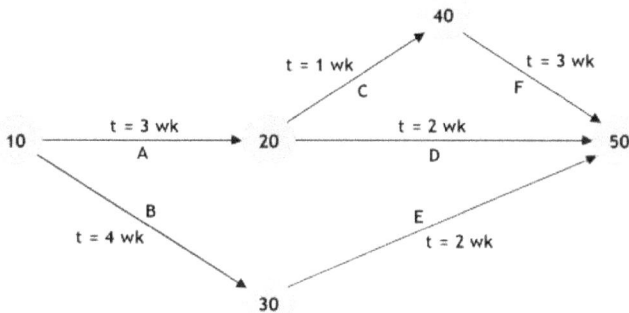

Figure 7.4: PERT Network.

Milestones are numbered so that the end node of an activity has a higher number than the start node. Incrementing the numbers by 10 allows for additional nodes to be inserted without modifying the numbering of the entire network. The activities are labelled alphabetically, and the expected time required for each activity is also indicated. The critical path is the pathway through the project network that takes the longest to complete, and will determine the overall time required to complete the project. Bear in mind that for a complex project with many activities and task dependencies, there can be more than one critical path through the network, and that the critical path can change.

PERT planning involves the following steps:

> Locate activities and milestones - tasks required to complete the project, and the events that mark the beginning and end of each activity, are listed in a table.

> Determine proper sequence of activities - this step may be combined with step 1, if the order in which activities must be performed is relatively easy to determine.

> Construct network diagram - using the results of steps 1 and 2, a network diagram is drawn which shows activities as arrowed lines, and milestones as circles. Software packages are available that can automatically produce a network diagram from tabular information.

> Estimate time required for each activity - any consistent unit of time can be used, although days and weeks are a common.

> Determine critical path - the critical path is determined by adding the activity times for each sequence and determining the longest path in the project. If the activity time for activities in other paths is significantly extended, the critical path may change. The amount of time that a non-critical path activity can be extended without delaying the project is referred to as its slack time.

> Updating PERT chart as project progresses - as the project progresses, estimated times can be replaced with actual times.

Because the critical path determines the completion date of the project, the project can be completed earlier by allocating additional resources to the activities on the critical path. PERT also identifies activities that have slack time, and which can therefore lend resources to critical path activities. One drawback of the model is that if there is little experience in performing an activity, the activity time estimate may simply be a guess. Another more serious problem is that, because another path may become the critical path if one or more of its associated activities are delayed, PERT often tends to underestimate the to time required to complete the project.

PERT incorporates uncertainty by making it possible to schedule a project while not knowing precise details and durations of all activities. The time shown for each project activity when creating the network diagram is the time that the task is expected to take based on a range of possibilities that can be defined as:

> The optimistic time - the minimum time required to complete a task

> The pessimistic time - the maximum time required to complete a task

> The most likely time - an estimate of how long the task will actually take

The expected time is defined as the average time the task would require if it were repeated a number of times over a period of time, and can be calculated using the following formula:

expected time = (optimistic time + (4 x most likely time) + pessimistic time) / 6

The information included on network diagram for activity includes activity name, expected duration, earliest start (ES), earliest finish (EF), latest start (LS), latest finish (LF) and slack. To determine these parameters, project activities to be identified and expected duration of each needs to be calculated. The earliest start (ES) for any activity will depend on the maximum earliest finish (EF) of all predecessor activities. The earliest finish for the activity is the earliest start plus the expected duration. The latest start (LS) for an activity will be equal to the maximum earliest finish of all predecessor activities. The latest finish (LF) is the latest start plus the expected duration. The slack in any activity is defined as the difference between the earliest finish and the latest finish, and represents the amount of time that a task could be delayed without causing a delay in subsequent tasks or the project completion date. Activities on the critical path by definition have zero slack.

A PERT chart provides a realistic estimate of the time required to complete a project, identifies the activities on the critical path, and makes dependencies visible. It can also identify the earliest and latest start and finish dates for a task, and any slack available. Resources can thus be diverted from non-critical activities to those that lie on the critical path should the need arise, in order to prevent project slippage. Variance in the project completion time can be calculated by summing the variances in the completion times of the activities in the critical path, allowing the probability of the project being completed by a certain date to be determined. PERT charts can become unwieldy, however, if the number of tasks is too great. The accuracy of the task duration estimates will also depend on the experience and judgment of the individual or group that make them.

8 | Problems

Network Modeling

1. Find the shipping plan between sources and destinations that minimizes total shipping cost while meeting the demands. The numbers in the table body show unit shipping costs.

	D1	D2	D3	D4	Supplies
S1	10	10	6	15	10
S2	5	15	10	12	15
S3	11	8	7	21	8
Demands	5	3	8	17	

Solution

The network model is shown below. The heavy arcs carry flow in the optimum solution. The minimum cost solution has z = 327.

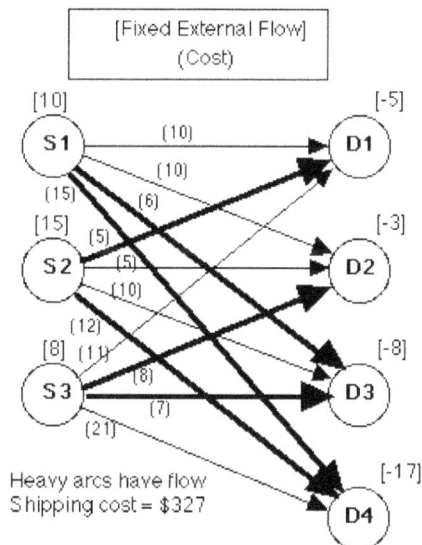

2. Shipping costs production data and demand data are given below. Find the production schedule which maximizes profit.

	D1	D2	D3	D4	Supplies
S1	10	10	6	15	10
S2	5	15	10	12	15
S3	11	8	7	21	8
Demands	5	3	8	17	

	Destination Data					Destination Data		
	Min Prod.	Max Prod	Cost per unit			Min Sakes.	Max Sales	Cost Revebyes
S1	10	15	$10	D1	D1	5	5	$20
S2	15	15	12	D2	D2	3	10	25
S3	0	8	13	D3	D3	0	8	22
				D4	D4	10	20	30

Solution

The network model is shown below. The optimum profit is 190.

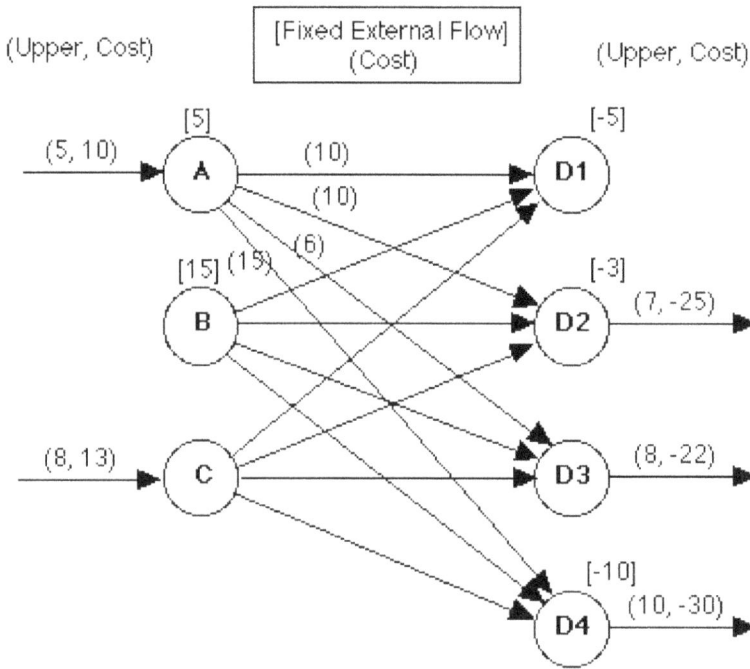

Optimum Profit 190.

3. A company has one manufacturing plant and three sales outlets (A, B, and C). Shipping costs from the plant to the outlets A, B, and C are respectively $4, $6, and $8. The company is to plan production, shipping and sales for three periods. The data for the periods is shown below. The plant has the capability to store some of the production of one period for sale in a later period. The storage cost is $1 per unit per period. The maximum storage is 100 units. Find a plan to maximize profit.

	Manufacturing Data Unit		Selling Price at out			Selling Price at out		
Period	Cost	Capacity	A	B	C	A	B	C
1	$8	175 units	$15	$20	$14	50	100	75
2	10	200	18	17	21	75	150	75
3	11	150	15	18	17	20	80	50

Solution

The network model for this problem has three subnetworks, each representing one of the three periods. The nodes labeled P1, P2, and P3 represent the manufacturing plant in the three periods. Arcs leaving these nodes carry the shipments to the outlets and are assigned the shipping costs. The external flows on these nodes describe the manufacturing costs and capacities in the three periods.

Inventories are carried on the arcs from P1 to P2 and P2 to P3. The parameters on these arcs indicate the inventory cost and capacity.

Outlets are modeled by the nodes labeled A, B, and C with maximum sales and revenues specified by the slack external flows.

Solution Profit = 1455.

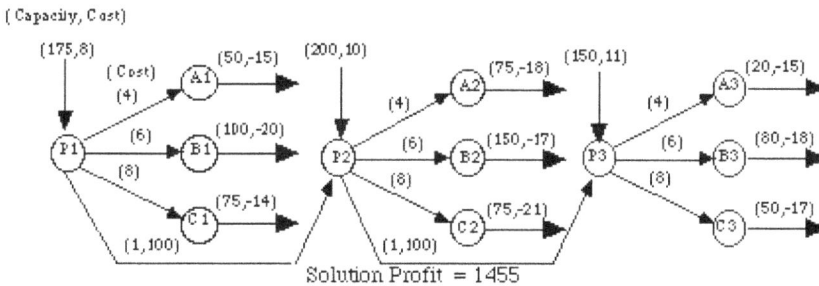

Solution Profit = 1455

Period	Prod.	Inv.	SalesA	SalesB	SalesC
1	175	25	50	100	0
2	200	0	75	75	75
3	100	0	20	80	0

4. Five crews are available to do three jobs. The table shows the time in hours required for each crew to do each job completely, but a crew can do part of a job in a corresponding fraction of the time. The table also shows the hourly wage of the crews. The crews are only paid for the time they work, and can work no more

than 20 hours. Find the assignment of jobs to crews that minimizes the total cost of completing the jobs.

	Crews				
Jobs	1	2	3	4	5
1	15	20	40	35	45
2	20	22	30	50	35
3	25	32	50	48	60
Cost per Hour	150	130	100	80	70

Solution

The Nodes on left represent crews. D and E are not shown. The arcs entering each crew node shows the maximum hours as an upper bound and the cost per hour as an arc cost. The nodes on the right represent jobs. The -1 fixed flow is the requirement that the job be done. The gain on arc (i,j) is proportion of the job completed in one hour, or the inverse of the time to complete the job.

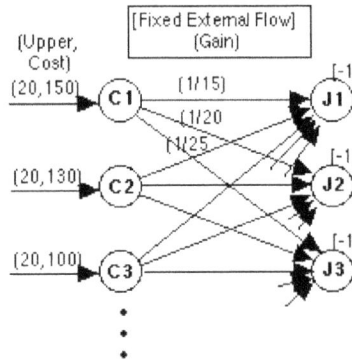

The table below shows the optimum distribution of crew hours to jobs. The table was computed using the Transportation model of the Mathematical programming add-in.

Net Cost: 8831

Trans. Flows		1	2	3	Supply Data		
	Name	J1	J2	J3	Min.	Max.	Cost
1	C1	15	0	5	0	20	150
2	C2	0	7.7333	12.267	0	20	130
3	C3	0	2.3117	0	0	20	100
4	C4	0	0	20	0	20	80
5	C5	0	20	0	0	20	70
Demand	Min.:	1	1	1			
Data	Max:	1	1	1			
	Revenue:	0	0	0			

5. Solve the assignment problem as a network flow problem.

	E	F	G	H
A	3	2	14	3
B	21	20	17	5
C	14	14	16	4
D	14	3	14	4

Solution

The network model is a bipartite network with nodes A, B, C and D as sources with external flows of +1. The destinations are E, F, G, and H with external flows of -1. The numbers in the matrix are the arc costs.

One optimum assignment is A to E, B to H, C to G and D to F.
An alternative optimum is A to E, B to G, C to H and D to F.
The objective is 27.

6. A company has three workers. On a particular work day, six jobs are scheduled to be completed. A cost is estimated for each worker and job and is shown in the table. Solve as a network problem.

a) Find the minimum cost assignment when each worker can do two jobs.

b) Each worker can do only one job. The primary goal is to finish as many jobs as possible. A secondary goal is to minimize cost.

c) Solve problem b when each worker can do any number of jobs.

Job

	1	2	3	4	5	6
A	3	2	2	6	4	6
B	4	3	7	5	3	3
C	9	9	7	9	7	6

Solution

The network model is a bipartite network with worker nodes A, B, and C as sources. The destinations are the job nodes 1 through 6. The numbers in the cost matrix are the arc costs from workers to jobs. The parts differ in how the flows enter or leave the network.

a) Fixed external flow of 2 for each worker node and -1 for each job node. The solution has: Assign A to jobs 2 and 3, B to jobs 1 and 5, and C to jobs 4 and 6. Cost is 26.

b) Fixed external flow of 1 for each worker node and and an arc leaving each job node with upper bound 1. The solution has: Assign A to jobs 3, B to job 5, and C to job 6. Cost less 300 is 11.

c) An arc entering each worker node with upper bound of 6, and each job node has an external flow of -1. The solution has: Assign A to jobs 1, 2 and 3, B to jobs 4, 5, and 6. Cost is 18.

7. Find the shortest path tree rooted at node 1 for the network shown.

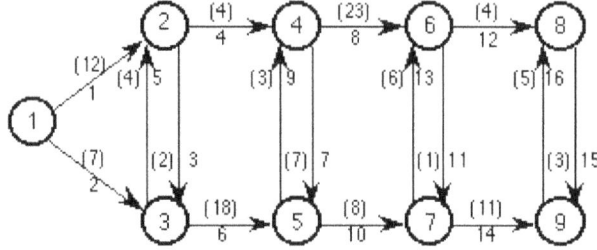

Solution

Set up the network model with an external flow of 8 on node 1 and external flow of -1 on the other nodes.

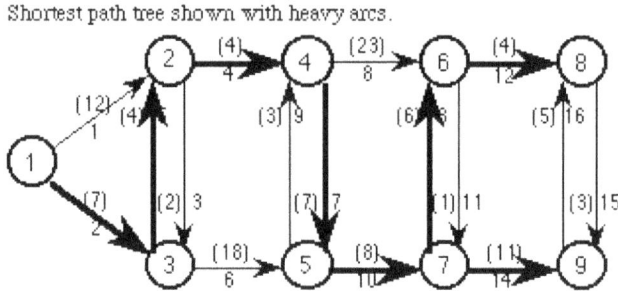

Shortest path tree shown with heavy arcs.

8. Assume the branches of the figure are all directed from left to right.

a) Let the numbers on the arcs represent branch lengths and find the shortest path tree with node 1 the origin node.

b) Let the numbers represent branch capacity and find the maximum flow with node 1 the source node and node 9 the sink node.

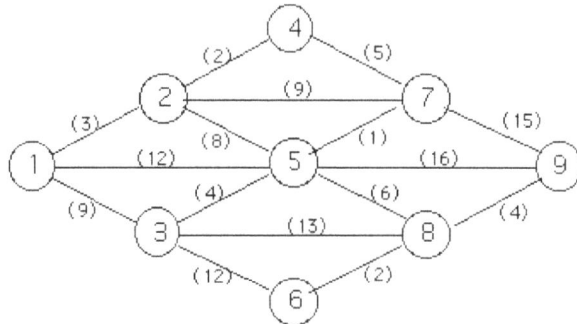

Solution

Part a. Set up the network model with an external flow of 8 on node 1 and external flow of -1 on the other nodes. Let the numbers on the arcs be the arc costs. Use default values for the other parameters.

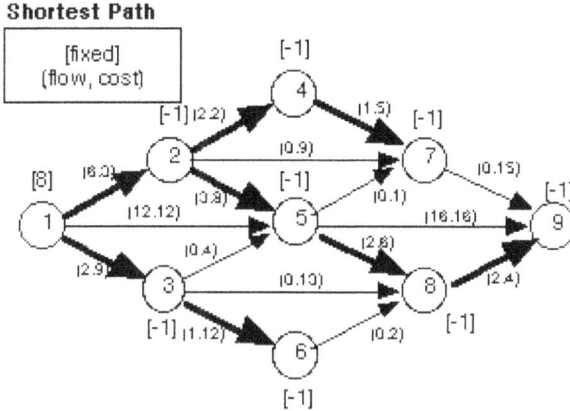

Part b. Set up the network model with an arc entering node 1 with a large upper bound. Have an arc leaving node 9 with a large upper bound and a cost of -1. Use the numbers given in the figure as the arc upper bounds on flow. Let lower bounds and costs be zero. All gains are 1.

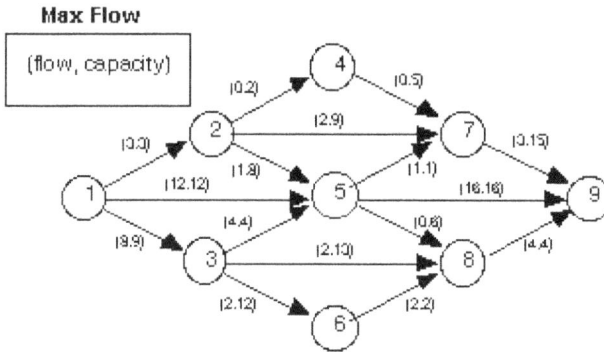

Bibliography

A. Bagchi, A. Chaudhary, and P. Kolman. Short length Menger's Theorem and reliable optical routing. Theoretical Computer Science, 339:315{332, 2005. Prelim. version in SPAA 03 (revue paper).

A. Baveja and A. Srinivasan. Approximating low-congestion routing and column-restricted packing problems. Information Processing Letters, 74:19{25, 2000.

A. Baveja and A. Srinivasan. Approximation algorithms for disjoint paths and related routing and packing problems. Mathematics of Operations Research, 25:255{280, 2000.

A. Beshir and F. A. Kuipers, "Variants of the min-sum link-disjoint paths problem," in IEEE Proc. Annu. Symp. Commun. & Vehicular Tech. (SCVT'09), 2009.

A. Bley, "On the complexity of vertex-disjoint lengthrestricted path problems," Computational Complexity, vol. 12, no. 3-4, pp. 131–149, 2003.

A. Chakrabarti, C. Chekuri, A. Gupta, and A. Kumar. Approximation algorithms for the unsplittable problem. In Proc. APPROX '02, pages 51{66. Springer-Verlag, 2002. C. Chekuri and S. Khanna. Edge disjoint paths revisited. In Proceedings of the 14th ACM-SIAM Symposium on Discrete Algorithms, pages 628{637, 2003.

A. Itai, Y. Perl, and Y. Shiloach, "The complexity of finding maximum disjoint paths with length constraints," Netw., vol. 12, no. 3, pp. 277–286, 1982.

A. Schrijver, "Finding k disjoint paths in a directed planar graph," SIAM J. Comput., vol. 23, no. 4, pp. 780–788, 1994.

A. Sprintson, M. Yannuzzi, A. Orda, and X. MasipBruin, "Reliable routing with QoS guarantees for multidomain IP/MPLS networks," in IEEE Conf. Comput. & Commun. (INFOCOM'07), 2007, pp. 1820–1828.

B. Awerbuch, Y. Azar, and S. Plotkin. Throughput-competitive online routing. In Proceedings of the 34th Annual IEEE Symposium on Foundations of Computer Science, pages 32{40, 1993.REFERENCES 20

Bedworth, David D. 1973. Industrial Systems: Planning, Analysis, and Control. New York: The Ronald Press Co.

Bodin, L. D., B. L. Golden, A. A. Assad, and M. O. Ball (1983), Computers and Operations Research, Special Issue on the Routing and Scheduling of Vehicles and Crews, Vol. 10, No. 2, pp. 63-211.

Brown, James Taylor. 1995. Priority Rule Search Technique for Resource Constrained Project Scheduling. Unpublished Ph.D. Dissertation, University of Central Florida.

C. Chekuri, S. Khanna, and B. Shepherd. Edge-disjoint paths in planar graphs. In Proceedings of the 45th Annual IEEE Symposium on Foundations of Computer Science, pages 71{80, 2004.

C. Chekuri, S. Khanna, and B. Shepherd. Multicommodity well-linked terminals and routing problems. In Proceedings of the 37th annual ACM Symposium on Theory of Computing, pages 183192, 2005.

C. Gao, M. M. Hasan, and J. P. Jue, "Domain-disjoint routing based on topology aggregation for survivable multidomain optical networks," OSA J. Opt. Commun & Netw, vol. 5, no. 12, pp. 1382–1390, 2013.

C.-L. Li, S. T. McCormick, and D. Simchi-Levi, "The complexity of finding two disjoint paths with minmax objective function," Discrete Applied Mathematics, vol. 26, no. 1, pp. 105–115, 1990.

Carlos Mart´ın-Vide, Gheorghe Paun, Juan Pazos & Alfonso Rodr ˘ ´ıguez-Paton (2003): ´ Tissue P systems. Theor. Comput. Sci. 296(2), pp. 295–326.

Casco, D. O., B. L. Golden, and E. A. Wasil (1988), "Vehicle Routing with Backhauls: Models, Algorithms, and Case Studies," in Golden and Assad (eds.), Vehicle Routing: Methods and Studies, pp 127-147.

Chancellor, A. (1988), "Snack Food Companies Haul More Than Chips," The Journal of Commerce and Commercial, March 21 edition, p. 1A.

Clarke, G. and J. Wright (1964), "Scheduling of Vehicles From a Central Depot to A Number of Delivery Points," Operations Research, Vol. 12, pp. 568-581.

Cooper, Dale. 1976. Heuristics for Scheduling Resource-Constrained Projects: An Experimental Investigation. Management Science, 22 (Jul.), 1186-1194.

Cullen, F. H. (1984), "Set Partitioning Based Heuristics For Interactive Routing," Unpublished Doctoral Dissertation, School of Industrial and Systems Engineering, Georgia Institute of Technology, Atlanta, Georgia.

Cullen, F. H., J. J. Jarvis, and H. D. Ratliff (1981), "Set Partitioning Based Heuristics for Interactive Routing," Networks, Vol. 11, No. 2, pp. 125-143.

D. A. Castanon, "Efficient algorithms for finding the k best paths through a trellis," IEEE Trans. Aerospace & Electronic Syst., vol. 26, no. 2, pp. 405–410, 1990.

D. A. Dunn, W. D. Grover, and M. H. MacGregor, "Comparison of k-shortest paths and maximum flow routing for network facility restoration," IEEE J. Selec. Areas in Commun., vol. 12, no. 1, pp. 88–99, 1994.

D. Bienstock and G. Iyengar. Solving fractional packing problems in (1=") iterations. In Proceedings of the 36th Annual ACM Symposium on Theory of Computing, pages 146{155, 2004.

D. Bienstock and M. A. Langston. Algorithmic implications of the Graph Minor Theorem. In M. O. Ball, T. L. Magnanti, C. L. Monma, and G. L. Nemhauser, editors, Handbook in Operations Research and Management Science 7: Network models. North-Holland, 1995.

D. Coudert, P. Datta, S. Perennes, H. Rivano, and ´M.-E. Voge, "Shared risk resource group complexity and approximability issues," Parallel Processing Letters, vol. 17, no. 02, pp. 169–184, 2007.

D. Xu, Y. Chen, Y. Xiong, C. Qiao, and X. He, "On the complexity of and algorithms for finding the shortest path with a disjoint counterpart," IEEE/ACM Transactions on Netw., vol. 14, no. 1, pp. 147–158, 2006.

D. Xu, Y. Xiong, and C. Qiao, "A new PROMISE algorithm in networks with shared risk link groups," in IEEE Global Telecommunications Conf. (GLOBECOM'03), vol. 5, 2003, pp. 2536–2540.

D. Zhou and S. Subramaniam, "Survivability in optical networks," IEEE Netw., vol. 14, no. 6, pp. 16–23, 2000.

Daeho Seo & Mithuna Thottethodi (2009): Disjoint-path routing: Efficient communication for streaming applications. In: IPDPS, IEEE, pp. 1–12.

Davis, E.W, and Heidorn, G.E. 1971. Optimal Project Scheduling Under Multiple Resource Constraints. Management Science, 17 (Aug.), 803-816.

Davis, E.W, and Patterson, J.H. 1975. A Comparison of Heuristic and Optimum Solutions in Resource-Constrained Project Scheduling. Management Science, 21 (Apr.).

Davis, E.W. 1973. Project Scheduling Under Resource Constraints: Historical Review and Categorization of Procedures. AIIE Transactions (Dec.), 297-313.

Davis, E.W. 1974. Networks: Resource Allocation. Industrial Engineering (Apr.).

Deif, I. and L. Bodin (1984), "Extension of the Clarke and Wright Algorithm For Solving the Vehicle Routing Problem With Backhauling," Proceedings of the Babson Conference on Software Uses in Transportation and Logistics Management, A. E. Kidder, Editor, Babson Park, MA

Desrochers, M., (1988). "An Algorithm for the Shortest Path Problem with Resource Constraints", GERAD Technical Report G-88-27, Ecole des Hautes Etudes Commerciales, Montreal.

Desrochers, M., Desrosiers, J. and M. Solomon, (1992). "A New Optimization Algorithm for the Vehicle Routing Problem with Time Windows." Operations Research, Vol. 40, No. 2, pp. 342-354.

E. A. Dinitz. Algorithm for solution of a problem of maximum in networks with power estimation. Soviet Math. Dokl., 11:1277{1280, 1970.

E. W. Dijkstra, "A note on two problems in connexion with graphs," Numerische Mathematik, vol. 1, no. 1, pp. 269–271, 1959.

Elsayed, E.A. 1985. Analysis and Control of Production Systems. Englewood Cliffs, N.J.: Prentice Hall

F. A. Kuipers, "An overview of algorithms for network survivability," ISRN Commun. Netw., vol. 2012, p. 24, 2012.

F. Kammer and T. Tholey, "The k-disjoint paths problem on chordal graphs," in Graph-Theoretic Concepts in Comput. Sci. Springer, 2010, pp. 190–201.

Fisher, M. L. and R. Jaikumar (1978), "A Decomposition Algorithm for Large-scale Vehicle Routing," Working Paper 78-11-05, Wharton Department of Decision Sciences, University of Pennsylvania.

Fisher, M. L. and R. Jaikumar (1981), "A Generalized Assignment Heuristic for Vehicle Routing," Networks, Vol. 11, No. 2, pp. 109-124.

Fisher, M. L., R. Jaikumar, and L. Van Wassenhove (1986), "A Multiplier Adjustment Method for the Generalized Assignment Problem," Management Science, Vol. 32, No. 9, pp. 1095-1103.

Francis, R. L. and J. A. White (1974), Facility Layout and Location, Prentice-Hall Inc., Englewood Cliffs, New Jersey.

G. Baier, E. Kohler, and M. Skutella. On the k-splittable flow problem. Algorithmica, 42:231{248, 2005. Special issue on ESA 2002.

Gelinas, S. and M. Desrochers, (1991), Personal communication.

Gelinas, S., (1991). "Fabrication de Tournees avec Rechargement" (in French, Construction of Backhaul Routes). Unpublished Master's Thesis, Ecole Polytechnique, Montreal.

Gheorghe Paun (1998): ˘ Computing with Membranes. Technical Report 208, Turku Center for Computer Science-TUCS. (www.tucs.fi).

Gheorghe Paun (2002): ˘ Membrane Computing: An Introduction. Springer-Verlag New York, Inc., Secaucus, NJ, USA.

Gheorghe Paun (2006): ˘ Introduction to Membrane Computing. In: Gabriel Ciobanu, Mario J. Perez-Jim´enez´ & Gheorghe Paun, editors: ˘ Applications of Membrane Computing, Natural Computing Series, Springer, pp.

Gillett, B. and L. Miller (1974), "A Heuristic Algorithm For the Vehicle Dispatch Problem," Operations Research, Vol. 22, pp.340-349.

Goetschalckx, M. and C. Jacobs-Blecha (1989), "The Vehicle Routing Problem With Backhauls," European Journal of Operational Research, Vol. 42, pp. 39-51.

Golden, B. L. and A. A. Assad Eds., (1988). Vehicle Routing: Methods and Studies. Elsevier Science Publishers, New York, New York.

Golden, B. L., E. K. Baker, J. L. Alfaro, and J. R. Schaffer (1985), "The Vehicle Routing Problem With Backhauling: Two Approaches," Working Paper Series MS/S 85-037, College of Business and Management, University of Maryland.

Golden, B., L. Bodin, T. Doyle, and W. Stewart, Jr. (1980), "Approximate Traveling Salesman Algorithms," Operations Research, Vol. 28, No. 3, Part II, pp. 694-711.

Haimovich, M., and A. H. G. Rinnooy Kan (1985), "Bounds and Heuristics for Capacitated Routing Problems," Mathematics of Operations Research, Vol. 10, No. 4, pp. 527-542.

Holloway, Charles A., Nelson, Rosser T., and Suraphongschai, Vicht. 1979. Comparison of a MultiPass Heuristic Decomposition Procedure with Other Resource-Constrained Project Scheduling Procedures. Management Science, 25 (Sept.), 862-872.

Hurtado, editors: Brainstorming Week on Membrane Computing, 2, Universidad de Sevilla, pp. 85–107.

J. Bermond, D. Coudert, G. D'Angelo, and F. Z. Moataz, "SRLG-diverse routing with the star property," in Int. Conf. Design of Reliable Commun. Netw. (DRCN'13), 2013, pp. 163–170.

J. Chuzhoy and S. Naor. New hardness results for congestion minimization and machine scheduling. In Proceedings of the 36th Annual ACM Symposium on Theory of Computing, pages 28{34, 2004.

J. Q. Hu, "Diverse routing in optical mesh networks," IEEE Trans. Commun., vol. 51, no. 3, pp. 489–494, 2003.

J. Suurballe, "Disjoint paths in a network," Netw., vol. 4, no. 2, pp. 125–145, 1974.

J. W. Suurballe and R. E. Tarjan, "A quick method for finding shortest pairs of disjoint paths," Netw., vol. 14, no. 2, pp. 325–336, 1984.

Jack Edmonds & Richard M. Karp (1972): Theoretical Improvements in Algorithmic Efficiency for Network Flow Problems. J. ACM 19(2), pp. 248–264.

Jacobs, C. D. (1987), "The Vehicle Routing Problem With Backhauls," Unpublished Doctoral Dissertation, School of Industrial and Systems Engineering, Georgia Institute of Technology, Atlanta, Georgia.

John T. Robacker (1956): Min-Max Theorems on Shortest Chains and Disjoint Cuts of a Network. Research Memorandum RM-1660, The RAND Corporation, Santa Monica, California.

Johnson, Roger V. 1992. Resource Constrained Scheduling Capabilities of Commercial Project Management Software. Project Management Journal, 22 (Dec.), 39-43.

Jordan, W. C. (1987), "Truck Backhauling on Networks With Many Terminals," Transportation Research-B, Vol. 21B, No. 3, pp. 183-193.

Jordan, W. C. and L. D. Burns (1984), "Truck Backhauling on Two Terminal Networks," Transportation Research-B, Vol. 18B, No. 6, pp. 487-503.

K. Kawarabayashi, Y. Kobayashi, and B. Reed, "The disjoint paths problem in quadratic time," J. Combinatorial Theory, Series B, vol. 102, no. 2, pp. 424–435, 2012.

Karl Menger (1927): Zur allgemeinen Kurventheorie. Fundamenta Mathematicae 10, pp. 96–115.

Kearney, A. T., Inc. (1984), Measuring and Improving Productivity in Physical Distribution, A report prepared for the National Council of Physical Distribution Management, NCPDM, Oak Brook, IL.

Khattab, M., and Choobineh, F. 1991. New Heuristic for Project Scheduling with a Single Resource Constraint. Computers and Industrial Engineering, 381-387.

L. Ford and D. R. Fulkerson, Flows in networks. Princeton Princeton University Press, 1962.

L. Guo and H. Shen, "On finding min-min disjoint paths," Algorithmica, vol. 66, no. 3, pp. 641–653, 2013.

Lenstra, J. K. and A. H. G. Rinnooy Kan (1981), "Complexity of Vehicle Routing and Scheduling Problems," Networks, Vol. 11, No. 2, pp. 221-227.

Lester R. Ford Jr. & D. Ray Fulkerson (1956): Maximal flow through a network. Canadian Journal of Mathematics 8, pp. 399–404.

Lin S., and B. Kernighan (1973), "An Effective Heuristic Algorithm for the Traveling Salesman Problem," Operations Research, Vol. 21, pp. 498-516.

M. Andrews and L. Zhang. Hardness of the undirected congestion minimization problem. In Proceedings of the 37th annual ACM Symposium on Theory of Computing, pages 284{293, 2005.

M. Andrews and L. Zhang. Hardness of the undirected edge-disjoint paths problem. In Proceedings of the 37th annual ACM Symposium on Theory of Computing, pages 276{283, 2005.

M. Andrews, J. Chuzhoy, S. Khanna, and L. Zhang. Hardness of undirected edge disjoint paths with congestion. In Proceedings of the 46th Annual IEEE Symposium on Foundations of Computer Science, pages 226{244, 2005.

M. C. Costa, L. Letocart, and F. Roupin. A greedy algorithm for multicut and integral multi ow in rooted trees. Operations Research Letters, 31:21{27, 2003.

M. C. Costa, L. Letocart, and F. Roupin. Minimal multicut and maximum integer multi ow: a survey. European Journal of Operational Research, 162:55{69, 2005.

M. Clouqueur and W. D. Grover, "Availability analysis of span-restorable mesh networks," IEEE J. Selec. Areas in Commun., vol. 20, no. 4, pp. 810–821, 2002.

M. J. Rostami, A. A. E. Zarandi, and S. M. Hoseininasab, "MSDP with ACO: A maximal SRLG disjoint routing algorithm based on ant colony optimization," J. Netw. Comput. Applications, vol. 35, no. 1, pp. 394–402, 2012.

M. T. Omran, J.-R. Sack, and H. Zarrabi-Zadeh, "Finding paths with minimum shared edges," J. Combinatorial Opt., vol. 26, no. 4, pp. 709–722, 2013.

M. To and P. Neusy, "Unavailability analysis of longhaul networks," IEEE J. Selec. Areas in Commun., vol. 12, no. 1, pp. 100–109, 1994.

Martello, S. and P. Toth (1981), "An Algorithm for the Generalized Assignment Problem," Operational Research '81, J. P. Brans, Ed., North-Holland Publishing Co.

Michael J. Dinneen, Yun-Bum Kim & Radu Nicolescu (2009): New Solutions to the Firing Squad Synchronization Problem for Neural and Hyperdag P Systems. In: Membrane Computing and Biologically Inspired Process Calculi, Third Workshop, MeCBIC 2009, Bologna, Italy, September 5, 2009, pp. 117–130.

Michael J. Dinneen, Yun-Bum Kim & Radu Nicolescu (2010): A Faster P Solution for the Byzantine Agreement Problem. In: Eleventh International Conference on Membrane Computing, CMC11, Jena, Germany, August 24-27, 2010, p. 26pp.

Michael J. Dinneen, Yun-Bum Kim & Radu Nicolescu (2010): P systems and the Byzantine agreement. Journal of Logic and Algebraic Programming 79(6), pp. 334 – 349.

Min, H., J. Current, and D. Schilling (1992), "The Multiple Depot Vehicle Routing Problem With Backhauling," Journal of Business Logistics, Vol. 13, No. 1, pp. 259-288.

Mittal, A. K., and V. Palsule (1984), "Facilities Location with Ring Radial Distances," IIE Transactions, Vol. 16, No. 1, pp. 59-64.

Moder, Joseph J., Phillips, Cecil R., and Davis, Edward W. 1983. Project Management with CPM, PERT and Precedence Diagramming. New York: Van Nostrand Reinhold.

N. Alon and J. Spencer. The Probabilistic method, 2nd edition. John Wiley and Sons, 2000.

N. Robertson and P. D. Seymour, "Graph minors. XIII. the disjoint paths problem," J. Combinatorial Theory, Series B, vol. 63, no. 1, pp. 65–110, 1995.

N. Taft-Plotkin, B. Bellur, and R. Ogier, "Quality-ofservice routing using maximally disjoint paths," in IEEE Int. Workshop on Quality of Service (IWQoS'99), 1999, pp. 119–128.

Nancy A. Lynch (1996): Distributed Algorithms. Morgan Kaufmann Publishers Inc., San Francisco, CA, USA.

Neil Robertson & Paul D. Seymour (1995): Graph minors. XIII: The disjoint paths problem. J. Comb. Theory Ser. B 63(1), pp. 65–110.

Orr, A. (1989), "Spartan Looks Down the Road," U. S. Distribution Journal, February, pp. 43-48.

P. Datta and A. K. Somani, "Graph transformation approaches for diverse routing in shared risk resource group (SRRG) failures," Comput. Netw., vol. 52, no. 12, pp. 2381–2394, 2008.

P. Erd}os and J. L. Selfridge. On a combinatorial game. Journal of Combinatorial Theory A, 14:298{301,1973. Bartholdi, J. J., L. K. Platzman, R. L. Collins, and W. H. Warden (1983), "A Minimal Technology Routing System for Meals-on-Wheels," Interfaces, Vol. 13, No. 3, pp. 1-8.

P. K. Agarwal, A. Efrat, S. K. Ganjugunte, D. Hay, S. Sankararaman, and G. Zussman, "The resilience of WDM networks to probabilistic geographical failures," IEEE/ACM Trans. Netw., vol. 21, no. 5, 2013.

P. Sebos, J. Yates, G. Hjalmtysson, and A. Greenberg, "Auto-discovery of shared risk link groups," in OSA Opt. Fiber Commun. Conf., vol. 4, 2001.

P. Zhang and W. Zhao, "On the complexity and approximation of the min-sum and min-max disjoint paths problems," in Combinatorics, Algorithms, Probabilistic and Experimental Methodologies. Springer, 2007, pp. 70–81.

Pascoe, T.L. 1965. An Experimental Comparison of Heuristic Methods for Allocating Resources. Unpublished Ph.D. Thesis, Cambridge University.

Patterson, James H. 1984. Comparison of Exact Approaches for Solving the Multiple Constrained Resource Project Scheduling Problem. Management Science, 30 (Jul.), 854-867.

R. Bellman, "On a routing problem," Quarterly of Applied Mathematics, vol. 16, pp. 87–90, 1958.

R. Bhandari, "Optimal diverse routing in telecommunication fiber networks," in IEEE Proc. Int. Conf. Comput. Commun. (INFOCOM'94), 1994, pp. 1498–1508.

R. Bhandari, "Optimal physical diversity algorithms and survivable networks," in IEEE Proc. Symp. Comput. & Commun., 1997, pp. 433–441.

R. K. Ahuja, T. L. Magnanti, and J. B. Orlin, Network flows: theory, algorithms, and applications. Prentice hall, 1993.

R. M. Karp, "On the complexity of combinatorial problems," Netw., vol. 5, pp. 45–68, 1975.

Radu Nicolescu, Michael J. Dinneen & Yun-Bum Kim (2009): Discovering the Membrane Topology of Hyperdag P Systems. In: Gheorghe Paun, Mario J. P ˘ erez-Jim ´ enez, Agust ´ ´ ın Riscos-Nu´nez, Grzegorz Rozenberg ˜ & Arto Salomaa, editors: Workshop on Membrane Computing, Lecture Notes in Computer Science 5957, Springer-Verlag, pp. 410–435.

Radu Nicolescu, Michael J. Dinneen & Yun-Bum Kim (2009): Structured Modelling with Hyperdag P Systems: Part A. In: Rosa Gutierrez-Escudero, Miguel A. Guti ´ errez-Naranjo, Gheorghe P ´ aun & Ignacio P ˘ erez- ´

Radu Nicolescu, Michael J. Dinneen & Yun-Bum Kim (2010): Towards Structured Modelling with Hyperdag P Systems. International Journal of Computers, Communications and Control 2, pp. 209–222.

S. Cosares and I. Saniee. An optimization problem related balancing loads on sonet rings. Telecommunications Systems, 3:165{181, 1994. Prelim. version as Technical Memorandum. Bellcore, Morristown, NJ, 1992.

S. D. Nikolopoulos, A. Pitsillides, and D. Tipper, "Addressing network survivability issues by finding the kbest paths through a trellis graph," in IEEE Conf. Comput. & Commun. (INFOCOM'97), vol. 1, 1997, pp. 370–377.

S. Fortune, J. Hopcroft, and J. Wyllie, "The directed subgraph homeomorphism problem," Theoretical Comput. Sci., vol. 10, no. 2, pp. 111–121, 1980.

S. Neumayer, A. Efrat, and E. Modiano, "Geographic max-flow and min-cut under a circular disk failure model," in IEEE Conf. Comput. & Commun. (INFOCOM'12), 2012, pp. 2736–2740.

S. Neumayer, G. Zussman, R. Cohen, and E. Modiano, "Assessing the vulnerability of the fiber infrastructure to disasters," IEEE/ACM Trans. Netw., vol. 19, no. 6, pp. 1610–1623, 2011.

S. Trajanovski, F. A. Kuipers, A. Ilic, J. Crowcroft, and P. Van Mieghem, "Finding critical regions and region disjoint paths in a network," IEEE/ACM Trans. Netw., vol. 23, no. 3, 2015.

S. Uludag, K.-S. Lui, K. Nahrstedt, and G. Brewster, "Analysis of topology aggregation techniques for QoS routing," ACM Comput. Surveys, vol. 39, no. 3, p. 7, 2007.

S. Yang, S. Trajanovski, and F. A. Kuipers, "Availabilitybased path selection," in Proc. Int. Workshop on Reliable Netw. Design & Modeling (RNDM'14), 2014.

S.-W. Lee and C.-S. Wu, "A k-best paths algorithm for highly reliable communication networks," IEICE Trans. Commun., vol. 82, no. 4, pp. 586–590, 1999.

Seibert, James E., and Evans, Gerald W 1991. Time-Constrained Resource Leveling. Journal of Construction Engineering and Management, 117, 3 (September), 503-520.

Stinson, Joel P., Davis, E.W, and Khumawala, B. 1978. Multiple Resource-Constrained Scheduling Using Branch and Bound. AIIE Transactions (Sept.).

T. Fenner, O. Lachish, and A. Popa, "Min-sum 2-paths problems," Lecture Notes in Comput. Sci., pp. 1–11, 2014.

Talbot, F. Brian, and Patterson, James H. 1976. An Efficient Integer Programming Algorithm with Network Cuts for Solving Resource-Constrained Scheduling Problems. Management Science, 24 (Dec.), 412-422.

U. Brandes, G. Neyer, and D. Wagner, "Edge-disjoint paths in planar graphs with short total length," Konstanzer Schriften in Mathematik und Informatik, vol. 19, 1996.

W. C. Lee, "Topology aggregation for hierarchical routing in ATM networks," Comp. Commun. Rev., vol. 25, no. 2, pp. 82–92, 1995

Whitehouse, Gary E. ed. 1979. Project Management: IIE Microsoftware Norcross, GA: Industrial Engineering.

Y. Aumann and Y. Rabani. Improved bounds for all-optical routing. In Proceedings of the 6th ACM-SIAM Symposium on Discrete Algorithms, pages 567{576, 1995.

Y. Azar and O. Regev. Strongly polynomial algorithms for the unsplittable ow problem. In Proceedings of the 8th Conference on Integer Programming and Combinatorial Optimization, pages 15{29, 2001.

Y. Dinitz, N. Garg, and M. X. Goemans. On the single-source unsplittable of problem. Combina-torica, 19:1{25, 1999. Preliminary version in FOCS 98.

Y. Guo, F. Kuipers, and P. Van Mieghem, "Link-disjoint paths for reliable QoS routing," Int. J. Commun. Syst., vol. 16, no. 9, pp. 779–798, 2003.

Y. Kobayashi and C. Sommer, "On shortest disjoint paths in planar graphs," Discrete Optimization, vol. 7, no. 4, pp. 234–245, 2010.

Yano, C. A., T. J. Chan, L. Richter, T. Culter, K. G. Murty, and D. McGettigan (1987), "Vehicle Routing at Quality Stores," Interfaces, Vol. 17, No. 2, pp. 52-63.

Terminology

Access List: List kept by routers to control access to or from the router for a number of services.

Activity: It is a distinct task that needs to be performed as part of the project.

Address Mapping: Technique that allows different protocols to interoperate by translating addresses from one format to another.

Adjacency: Relationship formed between selected neighboringrouters and end nodes for the propose of exchangingrouting information.

Algorithm: A set of rules and decision structures for actions in a specifically defined set of circumstances.

Application Layer: Layer 7 of the OSI reference model. This layer provides services to application processes (such as electronic mail, file transfer, and terminal emulation) that are outside of the OSI model.

Arc: The line or path between two nodes in a network.

Arrow: It shows the direction of the activity.

Assignment Problem: A specific class of LP problems that involves determining the most efficient assignment of people to projects, salespeople to territories, contracts to bidders, jobs to machines, and so on.

Attribute: Configuration data that defines the characteristics of database objects such as the chassis, cards, ports, or virtual circuits of a particular device. Attributes might be preset or user-configurable.

Backward pass: It is the process of moving through a project from finish to start to determine the latest start and finish times for the activities of the project.

Balanced Problem: The condition under which total demand (at all destinations) is equal to total supply (at all sources).

Bitmap: A data structure that uses bits to represent the attributes of an object that is not character-based.

Bridge: A Data Link Layer device that limits traffic between two network segments by filtering the data between them based on hardware addresses.

Bus Topology: Linear LAN architecture in which transmissions from network stations propagate the length of the medium and are received by all other stations.

CLNS: Connectionless Network Service. OSI network layer service that does not require a circuit to be established before data is transmitted.

Connectivity: A term referring to the ability of a device to trade data and shareresources with other devices of a similar and dissimilar typethrough electronic means including serial and parallel connections, networking and telecommunications.

Crashing: Crashing an activity refers to taking on extra expenditures in order to reduce the duration of an activity below its expected duration. Crash point shows the time and cost when the activity is fully crashed.

Critical path: This is the path that has the longest length through the project. It is the shortest time that a project can conceivably be finished. If the slack is zero for an activity then it is on critical path. Similarly if slack is positive then the activity is not on the critical path.

Data Link Layer: Layer 2 of the OSI reference model. This layer provides reliable transit of data across a physical link. The data link layer is concerned with physical addressing, network topology, line discipline, error notification, ordered delivery of frames, and flow control.

Destination or Sink: A demand location in a transportation problem.

Dummy: It is inserted into the network to show a precedence relationship, but it does not represent any actual passage of time.

Earliest finish time of an activity: It is the time at which an activity will finish if there is no delays in the project.

Earliest start of an activity: It is the calendar time when an event can occur when all the predecessor events completed at the earliest possible times. Earliest start time for an activity is equal to the largest of the earliest finish times of its immediate predecessors.

Facility Location Analysis: An application of the transportation model to help a firm decide where to locate a new factory, warehouse, or other facility.

Fiber-optic Cable: Physical medium capable of conducting modulated light transmission. Compared with other transmission media, fiber-optic cable is more expensive, but is not susceptible to electromagnetic interference and is capable of higher data rates. Sometimes called optical fiber.

File system: Refers to the collection of system software routines that managesand accesses files located on a computer's storage volumes.

Forward pass: It is the process of moving through the project from start to finish time determining the earliest start and finish times for the activities of the project.

Heterogeneous Network: Network consisting of dissimilar devices that run dissimilar protocols and in many cases support dissimilar functions or applications.

Hub-and-spoke: A transportation system in which travel from one area to another is routed through a central point, or hub.

Immediate predecessors: These are the activities that must be completed by no later than the start time of the given activity.

Immediate successor: Given the immediate predecessor of an activity, this activity becomes the immediate successor of each of these immediate predecessors. If an immediate successor has a multiple of immediate predecessors, then all must be finished before an activity can begin.

Internetwork: Collection of networks interconnected by routers and other devices that functions generally) as a single network.

ITU-T: International Telecommunication Union Telecommunication Standardization Sector. International body that develops worldwide standards for telecommunications technologies. The ITU-T carries out the functions of the former CCITT. See also CCITT.

Jabber: 1. Error condition in which a network device continually transmits random, meaningless data onto the network. 2. In IEEE 802.3, a data packet whose length exceeds that prescribed in the standard.

Jack: The female connector.

Jacket: The protective outer covering of a computer or network cable.

JANET: Joint Academic Network. X.25 WAN connecting university and research institutions in the United Kingdom.

Jitter: The difference between a real signal and its ideal due to distortion.

Jumper: Electrical switch consisting of a number of pins and a connector that can be attached to the pins in a variety of different ways. Different circuits are created by attaching the connector to different pins.

JUNET: Japan UNIX Network. Nationwide, noncommercial network in Japan, designed to promote communication between Japanese and other researchers.

JvNCnet: John von Neumann Computer Network. Regional network, owned and operated by Global Enterprise Services, Inc., composed of T1 and slower serial links providing midlevel networking services to sites in the Northeastern United States.

Karn's Algorithm: Algorithm that improves round-trip time estimations by helping transport layer protocols distinguish between good and bad round-trip time samples.

KB: Kilobyte.

KBPS: A unit of measure used to describe the rate of data transmission equal to 1000 bits per second.

KByte: A unit of measure used to describe an amount of information equal to 1024 (210) bytes.

Keepalive Interval: Period of time between each keepalive message sent by a network device.

Keepalive Message: Message sent by one network device to inform another network device that the virtual circuit between the two is still alive.

Kermit: Very slow telecom data-transfer protocol developed at Columbia, and used primarily in VAX environments, although widely ported. Like any other telecom data-transfer protocol it's purpose is to break a data stream into blocks, and provide flow-control, error detection, and re-transmission on the transfer of the blocks. Much less efficient than Xmodem, Ymodem, or Zmodem.

Kludge: A word used to describe a solution to a problem that lacks elegance or that contains components for a purpose significantly different that their original design purpose.

Label Swapping: Routing algorithm used by APPN in which each router that a message passes through on its way to its destination independently determines the best path to the next router.

LAN Manager: Distributed NOS, developed by Microsoft, that supports a variety of protocols and platforms.

LAN Server: A general term used to describe a device that manages and allows the use of more than one kind of resource such as storage or file services, print services, communication services, data base services, *etc.*

LAN Switch: High-speed switch that forwards packets between data-link segments. Most LAN switches forward traffic based on MAC addresses. This variety of LAN switch is sometimes called a frame switch. LAN switches are often categorized according to the method they use to forward traffic: cut-through packet switching or store-and-forward packet switching. Multilayer switches are an intelligent subset of LAN switches. Compare with multilayer switch. See also cut-through packet switching and store and forward packet switching.

LAN: A communication infrastructure that supports data and resource sharing within a small area (<2 km diameter) that is completely contained on the premises of a single owner.

LAP: Link Access Protocol. Any protocol of the Data Link Layer, such as EtherTalk.

LAPB: Link Access Procedure, Balanced. Data link layer protocol in the X.25 protocol stack.

LAPD: Link Access Procedure on the D channel. ISDN data link layer protocol for the D channel. LAPD was derived from the LAPB protocol and is designed primarily to satisfy the signaling requirements of ISDN basic access. Defined by ITU-T Recommendations Q.920 and Q.921.

LAPM: Link Access Procedure for Modems. ARQ used by modems implementing the V.42 protocol for error correction. See also ARQ and V.42.

Laser: Light amplification by simulated emission of radiation. Analog transmission device in which a suitable active material is excited by an external stimulus to produce a narrow beam of coherent light that can be modulated into pulses to carry data. Networks based on laser technology are sometimes run over SONET.

LaserWriter: Any of a group of laser printers that uses PostScript as an imaging language and can communicate using AppleTalk protocols.

LAT: Local Area Transport. DECnet's method for communication between terminals and terminal servers. LAT cannot be routed.

Latency: In data transmission, the delay in transmission time that occurs while information remains in a device's buffered memory (such as a bridge or router) before it can be sent along its path.

Latest finish time of an activity: It is the latest time that the activity can be completed without delaying the subsequent events and completion of the project. Latest finish time of an activity is equal to the smallest of the latest start times of its immediate successors.

Latest start time of an activity: It is the latest time that the activity can start without delaying the subsequent events and completion of the project.

Layer: A term used to describe a group of communication functions and the protocols implemented to perform them as defined by a network standards organization, most often referring to a group of functions as described by the OSI 7-Layer Model designated by the ISO.

LCI: Logical channel identifier. See VCN.

LCN: Logical channel number. See VCN.

Leaf Internetwork: In a star topology, an internetwork whose sole access to other internetworks in the star is through a core router.

Learning Bridge: Bridge that performs MAC address learning to reduce traffic on the network. Learning bridges manage a database of MAC addresses and the interfaces associated with each address. See also MAC address learning.

Leased Line: Transmission line reserved by a communications carrier for the private use of a customer. A leased line is a type of dedicated line. See also dedicated line.

LEC: 1. LAN Emulation Client. Entity in an end system that performs data forwarding, address resolution, and other control functions for a single ES within a single ELAN. A LEC also provides a standard LAN service interface to any higher-layer entity that interfaces to the LEC. Each LEC is identified by a unique ATM address, and is associated with one or more

MAC addresses reachable through that ATM address. See also ELAN and LES.

LECS: LAN Emulation Configuration Server. Entity that assigns individual LANE clients to particular ELANs by directing them to the LES that corresponds to the ELAN. There is logically one LECS per administrative domain, and this serves all ELANs within that domain. See also ELAN.

LED: Light emitting diode. Semiconductor device that emits light produced by converting electrical energy. Status lights on hardware devices are typically LEDs.

LEN node: Low-entry networking node. In SNA, a PU 2.1 the supports LU protocols, but whose CP cannot communicate with other nodes. Because there is no CP-to-CP session between a LEN node and its NN, the LEN node must have a statically defined image of the APPN network.

LES: LAN Emulation Server. Entity that implements the control function for a particular ELAN. There is only one logical LES per ELAN, and it is identified by a unique ATM address. See also ELAN.

Level 1 Router: Device that routes traffic within a single DECnet or OSI area.

Level 2 Router: Device that routes traffic between DECnet or OSI areas. All Level 2 routers must form a contiguous network.

Line Code Type: One of a number of coding schemes used on serial lines to maintain data integrity and reliability. The line code type used is determined by the carrier service provider. See also AMI and HBD3.

Line Conditioning: Use of equipment on leased voice-grade channels to improve analog characteristics, thereby allowing higher transmission rates.

Line Driver: Inexpensive amplifier and signal converter that conditions digital signals to ensure reliable transmissions over extended distances.

Line of Sight: Characteristic of certain transmission systems, such as laser, microwave, and infrared systems, in which no obstructions in a direct path between transmitter and receiver can exist.

Line Turnaround: Time required to change data transmission direction on a telephone line.

Link State Routing Algorithm Algorithm: Routing algorithm in which each router broadcasts or multicasts information regarding the cost of reaching each of its neighbors to all nodes in the internetwork. Link state algorithms create a consistent view of the network and are therefore not prone to routing loops, but they achieve this at the cost of relatively greater computational difficulty and more widespread traffic (compared with distance vector routing algorithms). Compare with distance vector routing algorithm. See also Dijkstra's algorithm.

Link State Routing: A routing protocol that takes link loading and bandwidth when selecting between alternate routes. Example: OSPF.

Link: Network communications channel consisting of a circuit or transmission path and all related equipment between a sender and a receiver. Most often used to refer to a WAN connection. Sometimes referred to as a line or a transmission link.

Little-endian: Method of storing or transmitting data in which the least significant bit or byte is presented first. Compare with big-endian.

LM/X: LAN Manager for INIX. Monitors LAN devices in UNIX environments.

LMI: Local Management Interface. Set of enhancements to the basic Frame Relay specification. LMI includes support for a keepalive mechanism, which verifies that data is flowing; a multicast mechanism, which provides the network server with its local DLCI and the multicast DLCI; global addressing, which gives DLCIs global rather than local significance in Frame Relay networks; and a status mechanism, which provides an on-going status report on the 'DLC Is known to the switch. Known as LMT in ANSI terminology.

LNM: LAN Network Manager. SRB and Token Ring management package provided by IBM. Typically running on a PC, it monitors SRB and Token Ring devices and can pass alerts up in NetView.

Load Balancing: In routing, the ability of a router to distribute traffic over all its network ports that are the same distance from the destination address. Good load-balancing algorithms use both line speed and reliability information. Load balancing increases the utilization of network segments, thus increasing effective network bandwidth.

Local Bridge: Bridge that directly interconnects networks in the same geographic area.

Local Explorer Packet: Generated by an end system in an SRB network to find a host connected to the local ring. If the local explorer packet fails to find a local host, the end system produces either a spanning explorer packet or an all-routes explorer packet. See also all-routes explorer packet, explorer packet, and spanning explorer packet.

Local Loop: Line from the premises of a telephone subscriber to the telephone company CO.

Local Traffic: Filtering Process by which a bridge filters out (drops) frames whose source and destination MAC addresses are located on the same interface on the bridge, thus preventing unnecessary traffic from being forwarded across the bridge. Defined in the IEEE 802.1 standard. See also IEEE 802.1.

LocalTalk: A Data Link Layer protocol defined in Inside AppleTalk by Apple Computer that covers the transmission of data on twisted pair wire using CSMA/CA at 230.4 Kilobits per second.

Logical Channel: Nondedicated, packet-switched communications path between two or more network nodes. Packet switching allows many logical channels to exist simultaneously on a single physical channel.

Loop: Route where packets never reach their destination, but simply cycle repeatedly through a constant series of network nodes.

Loopback packet: A test packet sent by a network adapter with a destination address equal to the adapter's own hardware address. The purpose of this test is typically to establish that the adapter is connected to a network that is functional enough to support a data transmission.

Loopback test: A test packet sent by a network adapter with a destination address equal to the adapter's own hardware address. The purpose of this test is typically to establish that the adapter is connected to a network that is functional enough to support a data transmission.

Loss: The aggregate attenuation of a signal due to interaction with its environment.

Lossy: Characteristic of a network that is prone to lose packets when it becomes highly loaded.

LSP: Link State Packet. A packet broadcast by a link state router listing the router's neighbors.

LU6.2: An SNA communications protocol that establishes a peer-to-peer session between two processes.

MacIP: Network layer protocol that encapsulates IP packets in DDS or transmission over AppleTalk. MacIP also provides proxy ARP services.

Magnetic field: The area surrounding an electrically charged body in which an electromagnetic force can be detected.

MAN: metropolitan-area network. Network that spans a metropolitan area. Generally, a MAN spans a larger geographic area than a LAN, but a smaller geographic area that a WAN.

Managed Object: In network management, a network device that can be managed by a network management protocol.

Management Services: SNA functions distributed among network components to manage and control an SNA network.

Management: Management of OSI networks. Accounting management subsystems are responsible for collecting network data relating to resource usage.

Manchester encoding: Digital coding scheme, used by IEEE 802.3 and Ethernet, in which a mid-bit-time transition is used for clocking, and a 1 is denoted by a high level during the first half of the bit time.

MAP: Manufacturing Automation Protocol. Network architecture created by General Motors to satisfy the specific needs of the factory floor. MAP specifies a token-passing LAN similar to IEEE 802.4.

MAPI: Microsoft Application Programming Interface. A programming library for Windows developers that provides messaging services to their applications.

MAU: Media Access Unit. The component of a network adapter that directly attaches to the Transmission media.

Maximal-Flow Problem: A problem that finds the maximum flow of any quantity or substance through a network.

Maximum Burst: Specifies the largest burst of data above the insured rate that will be allowed temporarily on an ATM PVC but will not be dropped at the edge by the traffic policing function, even if it exceeds the maximum rate. This amount of traffic will be allowed only temporarily; on average, the traffic source needs to be within the maximum rate. Specified in bytes or cells.

Maximum Rate: Maximum total data throughput allowed on a given virtual circuit, equal to the sum of the insured and uninsured traffic from the traffic source. The uninsured data might be dropped if the network becomes congested. The maximum rate, which cannot exceed the media rate, represents the highest data throughput the virtual circuit will ever deliver, measured in bits or cells per second.

MBONE: Multicast backbone. The multicast backbone of the Internet. MBone is a virtual multicast network composed of multicast LANs and the point-to-point tunnels that interconnect them.

MBPS: A unit of measure used to describe the rate of data transmission equal to one millions bits per second.

MByte: A unit of measure used to describe an amount of information equal to 1,048,576 (220) bytes.

MCA: Micro channel architecture. Bus interface commonly used in PCs and some UNIX workstations and servers.

MCI: Multiport Communications Interface. Card on the AGS+ that provides two Ethernet interfaces and up to two synchronous serial interfaces. The MCI processes packets rapidly, without the interframe delays typical of other Ethernet interfaces.

MCR: minimum cell rate. Parameter defined by the ATM Forum for ATM traffic management. MCR is defined only for ABR transmissions, and specifies the minimum value for the ACR.

MD5: Message Digest 5. Algorithm used for message authentication in SNMP v.2. MD5 verifies the integrity of the communication, authenticates the origin, and checks for timelines.

Media Rate: Maximum traffic throughput for a particular media type.

Media: The environment in which the transmission signal is carried.

Memory Allocation: The amount of memory, usually RAM, that an process reserves for itself.

Memory: In computing, a system where data is stored for direct, highspeed access by a microprocessor.

Mesh: Network topology in which devices are organized in a manageable, segmented manner with many, often redundant, interconnections strategically placed between network nodes.

Message Switching: Switching technique involving transmission of messages from node to node through a network. The message is stored at each node until such time as a forwarding path is available.

Message Unit: Unit of data processed by any network layer.

Message: Application layer (Layer 7) logical grouping of information, often composed of a number of lower-layer logical groupings such as packets. The terms datagram, frame, packet, and segment are also used to describe logical information groupings at various layers of the OSI reference model.

Metalanguage: A language that represents another language.

Metasignaling: Process running at the ATM layer that manages signaling types and virtual circuits.

MHS: 1. Message Handling Service. A synonym of X.400 store and forward messaging. 2. Message Handling System. A Novell protocol for mail handling.

MIB: Management Information Base. In SNMP, a specification of the data objects and data structures that the Agent is responsible for knowing and reporting to the Console on demand.

MIC: Media interface connector. FDDI de facto standard connector.

Micro: 1. A prefix that denotes a one millionth part of a unit of measure, such as a microsecond or microampere. 2. A prefix that denotes something small. 3. A slang term for any personal computer.

Microcode: Translation layer between machine instructions and the elementary operations of a computer. Microcode is stored in ROM and allows the addition of new machine instructions without requiring that they be designed into electronic circuits when new instructions are needed.

Microsegmentation: Division of a network into smaller segments, usually with the intention of increasing aggregate bandwidth to network devices.

Microwave: 1. Any electromagnetic radiation with a wavelength between 1 millimeter and 1 meter. 2. A point-to-point data transmission system employing electromagnetic radiation using a carrier frequency in the microwave region.

Midsplit: Broadband cable system in which the available frequencies are split into two groups: one for transmission and one for reception.

Minimal-Spanning Tree Model: Determines the path through the network that connects all of the nodes while minimizing total distance.

Mips: Millions of instructions per second. Number of instructions executed by a processor per second.

MIS: Management Information System. Used to describe the set of computing resources that hold and allow access to the information owned by an organization.

Mode: 1. One particular method or way of accomplishing a goal. 2. In fiber optic transmission, a particular path between a light source and a receiver. 3. In statistics, the result with the highest frequency within the sample group.

Modem: A device that can covert data signals between analog and digital signaling systems.

MOP: Maintenance Operation Protocol. Digital Equipment Corporation protocol that provides a way to perform primitive maintenance operations on DECnet systems.

MSAU: Multistation access unit. Wiring concentrator to which all end stations in a Token Ring network connect. The MSAU provides an interface between these devices and the Token Ring interface.

MTU: Maximum Transmission Unit. A specification in a data link protocol that defines the maximum number of bytes that can be carried in any one packet on that link.

Multiaccess Network: Network that allows multiple devices to connect and communicate simultaneously.

Multicast Group: Dynamically determined group of IP hosts identified by a single IP multicast address.

Multicast router: Router used to send IGMP query messages on their attached local networks. Host members of a multicast group respond to a query by sending IGMP reports noting the multicast groups to which they belong. The multicast router takes responsibility for forwarding multicast datagrams from one multicast group to all other networks that have members in the group.

Multicast Server: Establishes a one-to-many connection to each device in a VLAN, thus establishing a broadcast domain for each VLAN segment. The multicast server forwards incoming broadcasts only to the multicast address that maps to the broadcast address.

Multicast: Single packets copied by the network and sent to a specific subset of network addresses. These addresses are specified in the destination address field.

Multidrop Line: Communications line having multiple cable access points. Sometimes called a multipoint line.

Multihomed Host: Host attached to multiple physical network segments in an OSI CLNS network.

Multihoming: Addressing scheme in IS-IS routing that supports assignment of multiple area addresses.

Multilayer switch: Switch that filters and forwards packets based on MAC addresses and network addresses. A subset of LAN switch.

Multimode Fiber: Optical fiber supporting propagation of multiple frequencies of light.

Multitasking: A descriptive term for a computing device whose operating system can handle several tasks concurrently. In monoprocessors, each active task is given short periods of time to use the CPU in a rotational fashion.

Multi-user: A term used to describe a computing process that can handle the requirements of several users simultaneously.

Multivendor network: Network using equipment from more than one vendor.

MVS: Multiple Virtual Storage. The primary operating systems for IBM mainframes.

NAP: Network access point. Location for interconnection of internet service providers in the United States for the exchange of packets.

NAU: Network addressable unit. SNA term for an addressable entity. Examples include LUs, PUs, and SSCPs. NAUs generally provide upper-level network services.

NBP: Name Binding Protocol. The AppleTalk protocol that associates the name, type and zone of a process with its Internet Socket Address.

Net Flow: The difference between the total flow in to a node and the total flow out of the node.

NetBIOS: Network Basic Input/Output System. API used by applications on an IBM LAN to request services from lower-level network processes. These services might include session establishment and termination, and information transfer.

NetWare: A trademark of Novell that includes network operating systems and LAN server processes that run on and/or serve many computing platforms, operating systems and protocols.

Network Adapter: A hardware device that translates electronic signals between a computing device's native network hardware and the transmission media. A network adapter may also include memory or additional hardware or firmware to aid or perform the computing device's network operations.

Network Address: Network layer address referring to a logical, rather than a physical, network device. Also called a protocol address.

Network Administrator: A person who is charged with the responsibility of caring for a network and the communication abilities of its users.

Network Analyzer: Hardware or software device offering various network troubleshooting features, including protocol-specific packet decodes, specific preprogrammed troubleshooting tests, packet filtering, and packet transmission.

Network Architecture: A set of specifications that defines every aspect of a data network's communication system, including but not limited to the types of user interfaces employed, the networking protocols used and the structure and types of network cabling that may be used.

Network Flow Model: A special type of LP model, used for problems such as transportation and assignment, that deals with arcs (paths) connecting a number of nodes, or points.

Network Interface: Boundary between a carrier network and a privately-owned installation

Network Layer: Layer 3 of the OSI reference model. This layer provides connectivity and path selection between two end systems. The network layer is the layer at which routing occurs. Corresponds roughly with the path control layer of the SNA model.

Network Management: A set of activities and duties whose goal is to provide high-quality,reliable communication among a group of networked computerusers. Typical activities may include resource planning, network design, providing user assistance and training, reconfiguration ofthe network due to a change in user requirements, assessing userneeds and designing appropriate solutions and troubleshootingand remedying network problems as they arise.

Network Operator: Person who routinely monitors and controls a network, performing tasks such as reviewing and responding to traps, monitoring throughput, configuring new circuits, and resolving problems.

Network: A series of connections, such as transportation or telecommunications, between a number of nodes, or points.

Network: The infrastructure that supports electronic data exchange.

NFS: Network File System. A file metalanguage and set of procedurecalls to access and manage files that is standard issue on nearlyevery computer that uses TCP/IP protocols as its standard network protocols. Designed by Sun Microsystems, NFS is now astandard feature of nearly all Unix systems.

NHRP: Next Hop Resolution Protocol. Protocol used by routers to dynamically discover the MAC address of other routers and hosts connected to a NBMA network. These systems can then directly communicate without requiring traffic to use an intermediate hop, increasing performance in ATM, Frame Relay, SMDS, and X.25 environments.

NIS: Network Information System. Protocol developed by Sun Microsystems for the administration of network-wide databases. the service essentially uses two programs: one for finding a NIS server and one for accessing the NIS databases.

NMVT: Network management vector transport. SNA message consisting of a series of vectors conveying network management-specific information. NNNetwork node. SNA intermediate node that provides connectivity, directory services, route selection, intermediate session routing, data transport, and network management services to LEN nodes and ENs. The NN contains a CP that manages the resources of both the NN itself and those of the ENs and LEN nodes in its domain. NNs provide intermediate routing services by implementing the APPN PU 2.1 extensions.

NNI: Network-to-Network Interface. ATM Forum standard that defines the interface between two ATM switches that are both located in a private network or are both located in a public network. The interface between a public switch and private one is defined by the UNI standard. Also, the standard interface between two Frame Relay switches meeting the same criteria.

NOC: Network Operations Center. Organization responsible for maintaining a network.

Node: A networked computing device that takes a protocol address andcan initiate and respond to communication from other networked devices that employ similar protocols.

Node: A specific point or location in a network.

Node: It is represented by a circle and indicates an event, a point in time where one or more activities start and/or finish. Start node is that node that represents the beginning of the project while the finish node indicates the end of the project.

Noise: Undesirable electrical or electromagnetic signals.

Nonseed Router: In AppleTalk, a router that must first obtain and then verify its configuration with a seed router before it can begin operation.

Non-stub Area: Resource-intensive OSPF area that carries a default route, static routes, intra-area routes, interarea routes, and external routes. Nonstub areas are the only OSPF areas that can have virtual links configured across them and are the only areas that can contain an ASBR.

Non-Volatile: Information that will remain usable by a computer despite loss of power or shutdown.

Normal point: It is the time and cost of an activity when it is performed in a normal way.

NOS: Network operating system. Generic term used to refer to what are really distributed file systems.

Nubus: One of a large number of computer bus architecture's used in Macintosh computers.

Object: In NBP, the proper term for describing the individual name given to a service. In the Chooser, it is the name that is displayed in the list of devices.

ODI: Open Data-link Interface. A Novell specification for network interface card device drivers that allows multiple protocol stacks to use the same card simultaneously.

Packet: A discrete chunk of communication in a pre-defined format.

Path: A path through a project network is a route that follows a set of arcs from the start node to the finish node. The length of the path is defined as the sum of the durations of the activities of the path.

Peer: In networking, a device to which a computer has a network connection that is relatively symmetrical, *i.e.* where both devices can initiate or respond to a similar set of requests.

Physical Address: A synonym for Hardware Address or MAC – layer address.

Polling: A means of Media Access Control where a device may only transmit information when it is given permission to transmit by a controller device.

PPC: 1. Process-to-Process Communication. 2. Sometimes used as an acronym for PowerPC.

Precision: Referring to the smallest difference in measurement that a test instrument can distinguish.

Protocol: In networking, a specification of the data structures and algorithms necessary to accomplish a particular network function.

RAM: Random Access Memory. A group of memory locations that are numerically identified to allow high speed access by a CPU. In random access, any memory location can be accessed at any time by referring to its numerical identifier as compared to sequential access, where memory location 6 can only be accessed after accessing memory locations 1-5.

Response Time: The gap between the time when a user initiates an action and the time that the action displays its results.

RIP: Routing Information Protocol. A distance vector routing protocol for IP.

ROM: Read Only Memory. A chip or other electronic device that contains memory that cannot be altered. In the Macintosh, the ROM contains the Macintosh OS and instructions for basic system operations.

RPC: Remote Procedure Call. A command given by one computer to a second computer over a network to execute a defined system call, such as in an NFS session.

Session Layer: The layer in the OSI 7-Layer Model that is concerned with managing the resources required for the session between two computers.

Set: In SNMP, the command given by the Console that asks the MIB to change the value of a data object in its MIB.

Shortest-Path Problem: A problem that determines the shortest path or route through a network.

Signal: The means of conveyance for a communication, typically an electromagnetic wave that is modulated to encode the information communicated.

Slack time: It is the differences between the latest time and the earliest time an activity. It is the amount of time by which an activity can be delayed without delaying the completion of the project.

SMDS: Switched Multimegabit Data Service. A metropolitan area packet switching data network using T-1 and T-3 lines.

SNA: Systems Network Architecture. IBM's communications architecture and strategy.

SNMP: Simple Network Management Protocol. A de facto standard for management of networked devices using a simple request-response data retrieval mechanism.

Source or Origin: An origin or supply location in a transportation problem.

Source: The node or process transmitting information.

SPX: Sequential Packet Exchange. A Transport layer protocol developed by Novell to provide in-sequence data transfer.

Star Topology: LAN topology in which end points on a network are connected to a common central switch by point-to-point links. A ring topology that is organized as a star implements a unidirectional closed-loop star, instead of point-to-point links. Compare with bus topology, ring topology, and tree topology.

Statistical Multiplexing: Technique in which information from multiple logical channels can be transmitted across a single physical channel. Statistical multiplexing dynamically allocates bandwidth only to active input channels, making better use of available bandwidth and allowing more devices to be connected than with other multiplexing techniques. Also referred to as statistical time-division multiplexing or stat mux.

Stub Network: Network that has only a single connection to a router.

Switch: A switch is a device that forwards packets between nodes based on the packet's destination node address (either hardware or protocol), typically with a buffer time longer than a repeater but shorter than the transmission time of the packet.

Synchronization: Establishment of common timing between sender and receiver.

Synchronous: A communication system where stations may only transmit at prescribed intervals and must provide a timing pulse with their packet.

Tagged Traffic: ATM cells that have their CLP bit set to 1. If the network is congested, tagged traffic can be dropped to ensure deliver of higher-priority traffic. Sometimes called DE (discard eligible) traffic.

TCP: Transmission Control Protocol. A reliable Transport Layer Protocol for managing IP that supports re-transmission, sequencing and fragmentation.

TIC: Token Ring interface coupler. Controller through which an FEP connects to a Token Ring.

Token: Frame that contains control information. Possession of the token allows a network device to transmit data onto the network.

Topology: 1. The arrangement of computing devices in a network. 2. A term describing such an arrangement.

Transmission: The activity of sending or conveying information.

Transport: Any of the functions carried out by protocols in the Network or Transport Layers.

Transportation Model: A specific case of LP involving scheduling shipments from sources to destinations so that total shipping costs are minimized.

Transshipment Point: A point in a network that is both a source and a destination; flows go both in and out.

Transshipment Problem: An extension of the transportation problem in which some points have flows both in to and out of them.

Trap: In SNMP, a message sent from the Agent to the Console when the Agent detects that condition defined by the network manager has occurred.

Unbalanced Problem: A situation in which total demand is not equal to total supply.

Variable: 1) A situation or aspect that cannot be expressed explicitly because it may have one of several values. 2) In troubleshooting, an aspect of a problem that makes it differ from a normal situation. 3) In a mathematical expression, a symbol that represents a number.

WAN: Wide Area Network. A network that is created between and among devices separated by large distances (typically in excess of 50 miles).

Workgroup: A group of networked computer users who frequently communicate with each other and share common devices.

Index

www.ingramcontent.com/pod-product-compliance
Lightning Source LLC
Chambersburg PA
CBHW082006190326
41458CB00010B/3093